NEW YORK REVIEW BOOKS
CLASSICS

RELIGIO MEDICI *and* HYDRIOTAPHIA, OR URNE-BURIALL

SIR THOMAS BROWNE (1605–1682) was the son of a prosperous London merchant who died while his son was still young. Browne attended Winchester College and Oxford, then spent several years studying medicine at Montpellier, Padua, and Leiden, before receiving his MD in 1633. In 1637 he settled in Norwich where he practiced medicine and lived for the rest of his life. *Religio Medici* was first published in 1642, without the author's consent; a year later he approved a new printing (with some of the controversial material removed), and the book became a best seller, subsequently translated into several European languages (and placed on the Papal Index). Browne's eccentric encyclopedia, *Pseudodoxia Epidemica*, was first published in 1646 and went through six editions. His last work to be published in his lifetime, *Urne-Buriall*, appeared in 1658. Browne was knighted in 1671, when King Charles II, his queen, and his court came to Norwich.

STEPHEN GREENBLATT is the author of, among other books, *Will in the World: How Shakespeare Became Shakespeare* and *The Swerve: How the World Became Modern* (winner of the National Book Award and the Pulitzer Prize). He is the John Cogan University Professor of the Humanities at Harvard.

RAMIE TARGOFF is the author of *Common Prayer: The Language of Public Devotion in Early Modern England*; *John Donne: Body and Soul*; and the forthcoming *Posthumous Love: Eros and the Afterlife in Renaissance England*. She is the Jehuda Reinharz Director of the Mandel Center for the Humanities and a professor of English at Brandeis.

RELIGIO MEDICI
and
HYDRIOTAPHIA, OR URNE-BURIALL

SIR THOMAS BROWNE

Edited and with an introduction by
STEPHEN GREENBLATT *and*
RAMIE TARGOFF

NEW YORK REVIEW BOOKS

New York

THIS IS A NEW YORK REVIEW BOOK
PUBLISHED BY THE NEW YORK REVIEW OF BOOKS
435 Hudson Street, New York, NY 10014
www.nyrb.com

Library of Congress Cataloging-in-Publication Data
Browne, Thomas, Sir, 1605–1682.
Religio medici and Urne-burial / by Sir Thomas Browne ; edited and with an introduction by Stephen Greenblatt and Ramie Targoff.
p. cm. — (New York Review Books Classics)
ISBN 978-1-59017-488-3 (alk. paper)
1. Christian life. I. Greenblatt, Stephen, 1943– II. Targoff, Ramie. III. Title.
PR3327.A73 2012
828'.409—dc23
2011051523

ISBN 978-1-59017-488-3

Printed in the United States of America on acid-free paper.
10 9 8 7 6 5 4 3 2

CONTENTS

To
Thomas Laqueur
and
Adam Phillips

INTRODUCTION

1.

WHEN Samuel Taylor Coleridge praised Sir Thomas Browne for having "a brain with a twist,"[1] he captured one of the central reasons why it remains such a pleasure to read Browne's prose. The "twist" in Browne's intelligence—his idiosyncratic and often surprising ways of thinking—is matched only by the peculiarity of the topics he chooses to think about. Why write a formal treatise on the history of the quincunx in gardening, or the discovery of some ancient urns in a nearby field? Why take the time to ponder and refute the popular belief that diamonds are softened by the blood of goats, or that beavers bite off their testicles to escape hunters, or that Jews naturally stink? Why compile notes on elephants or bubbles or Icelandic flora and fauna, or assemble a list of plants that can be grafted? Why discourse upon the nature of cymbals used by the ancient Hebrews, or deliberate on the answers of the oracle of Apollo to Croesus, King of Lydia? Why opine on the kinds of fish that Jesus Christ ate with his disciples after his resurrection from the dead?[2]

In a certain sense, these topics are typical of the era in which Browne lived. This seventeenth-century physician, like so many educated men and women of his generation, possessed both great learning and expansive curiosity, and could confidently discourse upon nearly anything that came to his attention. Browne took very little at face value, and he scrutinized with the same intensity everything from fragments of bones to doctrinal beliefs. His restless mind, open to the world, refused to rely on received opinions, and yet his eccentric, unsystematic intelligence set him apart from those seventeenth-century scientists or philosophers who harnessed their wide-ranging interests to reason and rigorous logic.

Browne was a connoisseur of uncertainty who delighted in circuitous methods and ambiguous conclusions. Any given paragraph of Browne's prose is likely to contain several "but"s, "if"s, and "yet"s; it is also likely to contain several contrary opinions—"some think," "others say"—which are almost never reconciled. Browne relished the bounteousness of human thought, the untrammeled range and variety of its expression, and rarely wanted to forego this vastness for the narrow prospect of a single truth.

Browne's preference for multiple possibilities rather than concrete answers made him vulnerable to charges of amateurism and unseriousness as early as 1642, when Sir Kenelm Digby attacked *Religio Medici* for its skepticism, its penchant for "aequivocall considerations," and its failure to argue "scientifically and methodically."[3] It was as a writer and not as a medical scientist (nor as a botanist, classicist, archaeologist, or theologian, among other studies he pursued) that Browne ultimately distinguished himself. As a writer, his contributions to the English language and its literature are many, and even, in part, quantifiable. Browne was a great coiner of neologisms, finding English insufficiently expressive for all that he wanted to say. According to *The Oxford English Dictionary*, he was responsible for introducing more than one hundred words to the language, including the nouns "exhaustion," "hallucination," and "suicide"; the verbs "compensate," "invigorate," and "bisect"; the adjectives "precocious," "medical," and "literary." Browne's fluency in Latin and the ease with which he extended that language's reach into English shows on any page of his prose, when words like "diuturnity" (long duration) or "absumption" (wasting away) are used without any recognition of their foreignness or difficulty. (That neither of these Latinate coinages made it into English usage shows that his linguistic success had its limits.)

More than his contributions to the *OED*, Browne is celebrated for the quality of his prose, which was widely imitated and admired for centuries after his death. Samuel Johnson first described Browne's style at some length, although his assessment leaned more toward criticism than praise:

> It is vigorous, but rugged; it is learned, but pedantick; it is deep, but obscure; it strikes, but does not please; it commands, but does not allure: his tropes are harsh, and his combinations uncouth. He fell into an age, in which our language began to lose the stability which it had obtained in the time of Elizabeth; and was considered by every writer as a subject on which he might try his plastick skill, by moulding it according to his fancy. [4]

Johnson later conceded that Browne's "innovations are sometimes pleasing, and his temerities happy: he has many 'verba ardentia,' forcible expressions, which he would never have found, but by venturing to the utmost verge of propriety."

By contrast to Johnson, romantic writers like Charles Lamb, Coleridge, William Hazlitt, and Thomas De Quincey were unabashedly enthusiastic about Browne. They found in him a model for their own imaginative, often unconventional writing. De Quincey ecstatically described Browne's prose in *Urne-Buriall* as a "melodious ascent as of a prelude to some impassioned requiem breathing from the pomps of earth, and from the sanctities of the grave!"[5] What these authors savored was precisely what Johnson most disliked. Distaste for Browne's lack of measured judgment gave way to praise for his bold wrestling with unresolvable questions; disapproval of the irregularity of his style was replaced by amazement at the variety and intensity of his expressions. The labyrinthine sentences that Johnson castigated seemed, as the twentieth-century novelist W. G. Sebald would later put it admiringly, like "processions or a funeral cortège in their sheer ceremonial lavishness."[6] Browne was now celebrated as the master of eloquent and supremely musical prose.

Browne loved music, and in a remarkable sentence in *Religio*

Medici he marvels at its power over him: "it unties the ligaments of my frame, takes me to pieces, dilates me out of myself, and by degrees, mee thinkes, resolves me into Heaven." He went on to try to account for the deep significance of harmony, and in his exalted words of praise we can sense Browne's profoundest ambition for his own writing: "It is an Hieroglyphicall and shadowed lesson of the whole world, and the Creatures of God." Centuries of admiring readers have thrilled to this ambition and responded with delight to the unique harmonies of Browne's verbal music.

2.

Browne was born in October 1605 to Thomas Browne, a merchant in the Mercers' Company (a trade association in London), and Anne Garraway, his wife. Browne's father died in 1613, and his mother made a fiscally if not emotionally disastrous marriage to Sir Thomas Dutton, a courtier and military officer to whom she had already loaned money from the unsettled estate. After wasting much of the inheritance, Browne's mother and stepfather were forced to give over the executorship of the estate to a responsible uncle, Edward Browne, who managed things sensibly and put young Thomas on track for a proper gentleman's education. Browne entered Winchester in 1616 and then proceeded to Oxford, where he matriculated in 1623 at Broadgates Hall (now Pembroke College). It was at Oxford that he began his study of physic and anatomy, and after receiving both his B.A. (1627) and M.A. (1629), he went to the Continent to study medicine. He began his medical studies at Montpellier and then continued at Padua. The latter was renowned for its anatomy theater as well as its herb garden, two subjects that would preoccupy Browne throughout his long career. He took his medical degree from Leiden, where he matriculated in 1633.

Browne was a product of the seventeenth-century new learning, centrally engaged in medical or scientific discoveries. This engagement comes out clearly in a 1646 letter to a young doctor, Henry

Power, who was embarking on his medical studies. "Be sure you make yourself master of Dr. Harvey's piece *De Circul. Sang*," Browne advises, "which discovery I prefer to that of Columbus." Browne's preference for Harvey over Columbus reveals a characteristic fascination with what lies hidden beneath the surface—the burial mounds of Saxons and Danes, the ashes inside Roman urns, the circulation of the blood within the body—rather than with the earthly surfaces themselves.

Upon returning to England, Browne soon settled in Norwich, where he began his medical practice. In 1641, he wed Dorothy Mileham, to whom he was married for more than four decades. Before his marriage, he had written scathingly of sexual intercourse as an unfortunate burden: "I could be content that we might procreate like trees, without conjunction, or that there were any way to perpetuate the world without this triviall and vulgar way of coition." Whatever grafting techniques he may or may not have mastered, Browne fathered eleven children, of whom six lived to adulthood and four outlived their parents. Much of Browne's correspondence with his sons, Thomas and Edward, survives; the latter went on to become a physician himself, and Browne guided him throughout his career. The letters to his sons reveal on the whole an engaged and affectionate parent, as well as a liberal giver of advice in the tradition of *Hamlet*'s Polonius: "be carefull you eat very few grapes and fruits [in France]"; "put on a decent boldness and learn a good garb of body"; avoid reading too much Lucretius, "there being divers impieties in it, and 'tis no credit to be punctually versed in it"; "examine the spine of fishes & how the spinall marrowe is ordered in them"; "look upon Aristotle *de Animal.* upon all occasion." In a letter to Edward from 1676 filled with instructions for medical lectures his son was preparing, Browne included an unusual gift: "I have enclosed the ureters & *vesica* or bladder, such as it is, of carp which we had this day, butt I had a fayre one long ago & lost it." Browne was nothing if not a devoted father.

Sir Thomas Browne died on October 19, 1682, just one month shy of his seventy-seventh birthday. He was buried in the chancel of his local parish church in Norfolk, where his wife erected a memorial tablet in Latin and English, identifying him as the author of "Religio medici and other Learned books," and herself as his "affectionate wife 47 years." Browne lay in peace until 1840, when his skull was removed from the family vault after his coffin was damaged during preparations for another burial. The skull remained separated from the rest of his bodily remains and those of his family members until it was reburied in its former place in 1922. This seems a telling fate for a man who had written multiple treatises on human burials, and who had asked the question, "But who knows the fate of his bones, or how often he is to be buried?"

Browne's writings—essays, investigations, tracts, notes, observations, and letters—fill more than 1,500 pages in the standard modern edition. In this volume we present the two works that have been widely regarded as his greatest and most characteristic achievements. In a small compass these famous essays, *Religio Medici* and *Urne-Buriall*, encapsulate the gloriously eccentric qualities of mind and prose that constitute Browne's unique appeal.[7]

3.

The title of Browne's best-known work, *Religio Medici* (The Religion of a Doctor), was meant to have a paradoxical ring. *Ubi tres medici*, a proverbial saying went, *duo Athei*: Two out of three doctors are atheists. In his first sentence Browne acknowledges the imputation: "For my Religion . . . there be severall circumstances that might perswade the world I have none at all." The first of these circumstances is what he calls "the generall scandall of my profession," that is, the presumption that most physicians are skeptics. To this he conjoins a second circumstance: "the naturall course of my studies." Browne had pursued his medical training at distinguished universities reputed to encourage unusually free inquiry. At Montpellier, the pursuit of science was cordoned off from theology, so that it was possible to investigate natural causes without constantly invoking God as the first and final cause of all things. At Padua, where Giordano Bruno and other freethinkers had engaged in speculations condemned elsewhere as heretical, medical students were trained in anatomy, as they were in Leiden. The opening of the body—so vividly depicted by Browne's contemporary Rembrandt in *The Anatomy Lesson of Dr. Nicolaes Tulp*—was regarded by some as a transgression, a searching out of secrets God had intended to keep hidden within the envelope of the flesh. Dissection was, moralists warned, a step toward the impious conviction that the soul was mortal and that it decayed along with the flesh. But the medical professors under whom Browne studied regarded the practice as a necessary component in the education of any physician.

Browne cites a third reason for the world to assume that he had no religion at all: "the indifferency of my behaviour, and discourse in matters of Religion, neither violently defending one, nor with the common ardour of contention opposing another." "Indifferency" here does not mean that he failed to manifest any interest in matters of belief or that he neglected religious observance. It means that he failed to be swept up in the intensely partisan doctrinal currents that had steadily gathered force in the 1620s and '30s and were threatening to tear England apart.

According to his preface to the reader, Browne penned *Religio Medici* around 1635, when he returned to England after his three years of study abroad. He wrote it, he claimed, only for his own "private exercise and satisfaction," but a manuscript copy given to one friend led to the making and circulation of other copies. These copies suggest that Browne revised his essay sometime between 1638 and 1640, probably in response to the interest that it had provoked. That interest prompted the printing of an unauthorized, pirated edition in 1642. Nettled by this "most depraved copy," Browne was driven to bring out a corrected, authorized version, which appeared the following year.

It is difficult to gauge the truth of this account. There was a stigma attached to print in the sixteenth and seventeenth centuries, and socially sensitive authors often pretended that they had no choice but to let their work come out in public as opposed to more decorously restricted and private manuscript circulation. Still, Browne's assertion that his original "intention was not public" may be accurate. Though his work is not strictly private writing, such as the diaries kept by some of his contemporaries (most famously, Samuel Pepys), it is an extremely personal and idiosyncratic tapestry of philosophical musing, playful speculation, ethical inquiry, religious affirmation, and self-revelation.

Browne might well therefore have originally imagined a carefully chosen audience for so unusual an exercise. He may have been particularly uneasy about generally broadcasting his views on religion at a time when questions of church doctrine and belief were fiercely debated. But the publication of the pirated edition meant that he could no longer remain in the shadows.

In 1643, when the authorized version appeared in print, the "common ardour of contention" in matters of religion, from which Browne had kept his distance, had led to a full-scale civil war. At this point his "indifferency" was not strictly a personal matter: it was a political position, a refusal to side with those puritanical forces, allied with Parliament, that were fighting to overthrow the episcopacy and bring about what they believed to be a godly commonwealth.

In his preface, Browne was careful to nod sympathetically toward both sides: because of the same irresponsible printing practices from which he had suffered, he writes, he had lived to see "the name of his Majesty defamed, the honour of Parliament depraved." And in the work itself, he carefully revised the pirated edition in order to clarify his moderate adherence to Anglican orthodoxy and reduce his vulnerability to attacks from ideological extremists on either side. "There is no church wherein every point so squares unto my conscience," Browne writes, as if he had looked around to see which one he would freely choose, "whose articles, constitutions, and customes seeme so consonant unto reason, and as it were framed to my particular devotion, as this whereof I hold my beliefe, the Church of *England*." He regards himself as obliged to subscribe to whatever articles of faith that church ordains. But, he adds, "whatsoever is beyond, as points indifferent, I observe according to the rules of my private reason, or the humour and fashion of my devotion." Those "points indifferent"—what theologians called *adiaphora*, speculations that lie outside the strict circle of obligatory dogma—are in effect the subject of *Religio Medici*. At a time, as Browne wrote, in which "wee are beholding unto every one wee meete hee doth not kill us," it took considerable courage, as well as intellectual agility, to expose one's "private reason," the "humour and fashion" of one's devotion, and indeed one's inner self to the scrutiny of the general public.

As it happened, that public was larger than Browne probably anticipated: reprinted in 1645, 1648, 1656, and 1659, *Religio Medici* was not only widely read in England but also became something of an international sensation. Translated within three years of its publication into Latin, French, and twice into Dutch, it received one of the surest marks of European acclaim: it found its way onto the Roman *Index expurgatorius*—the Catholic "index" of banned books.

Though it was unusual to expose the workings of one's inner self to the world, it was not unprecedented: Michel de Montaigne had made such exposure the hallmark of his celebrated *Essays* that he saw into print in the 1580s, at a time of comparable religious violence in France. The *Essays* had a substantial impact in England, both in French and in an English translation by John Florio published in 1603. Most English readers of Montaigne responded to one or another of the themes to which the essayist returns again and again: the overwhelming power of custom, the limitations of human knowledge, the primacy of experience, the dangerous absurdity of fanatical belief, the vulgar pretensions of civilization, and the abiding truths of nature. Browne was also alert to these themes, but he seems to have grasped, more than any of his contemporaries, that Montaigne's greatest achievement was his representation of the ceaseless motions of a single mind.

Religio Medici was Browne's attempt, in the spirit of Montaigne, to capture something of his own mind's movements, to register his shifts of attention and affect, to display the tangled intellectual and religious materials from which his characteristic attitudes were fashioned, to make and unmake and remake his stance toward the world. The work is divided into parts and sections, but it is extremely difficult to extract from these divisions any organized scheme of inquiry. An idea is raised for consideration and pursued toward an eloquent conclusion, but the train of thought is interrupted by a digression, the digression leads to another topic, and this new topic introduces a different set of ruminations that are themselves the occasion for further digressions. After a series of zigzagging motions, Browne may return to the initial idea with renewed eloquence but with conclusions that often seem distinctly at variance with the ones toward which he seemed at first confidently to be heading.

Hence, for example, he remarks that we should take literally, as well as metaphorically, the biblical "*All flesh is grasse.*" After all, the creatures we consume are "but the hearbs of the field, digested into flesh in them" and that same grass is therefore "more remotely carnified in our selves." This leads to the thought that we are all, again quite literally, "*Antropophagi* and Cannibals," since our own flesh is made up of the matter that is endlessly recycled through grass and animals and the stomachs of our own ancestors. From here, after a brief look at cemeteries, those "dormitories of the dead," Browne laments death's "dismall conquest" of everything, only to shift direction and congratulate himself that he does not dote on life. On the contrary, so "abject" a view of existence does he have that without the prospect of death, he would be "the miserablest person extant." "No man ever desired life," Browne declares, "as I have sometimes death." Yet a few pages later—after a dizzying succession of reflections on the notion that Adam was thirty at his creation; on the womb as the "truest microcosme"; on the transmigration of silkworms as a mystical emblem of the afterlife; on his own innate bashfulness; on his sense that, though only thirty years old, he has outlived himself and begins, like Macbeth, "to bee weary of the Sunne"; on the theologians' idea that there shall be no gray hairs in heaven; on the tendency of age to turn "bad dispositions into worser habits"; on the reasons that some men die young and others live to advanced age—the current of Browne's thought has shifted: though he regards it as brave to despise death, he has decided that "where life is more terrible than death, it is then the truest valour to dare to live." A moment later, in a still greater shift, he writes that "Whosoever enjoyes not this life, I count him but an apparition."

These baffling changes in direction and outright contradictions did not seem to trouble Browne. On the contrary, he made a principle of them, as when he declares—in the midst, let us recall, of a civil war—that "I could never divide my selfe from any man upon the difference of an opinion, or be angry with his judgement for not agreeing with mee in that, from which perhaps within a few days I should dissent my selfe." The stance probably reflected wariness, but it also reflected a monumental egotism. To Browne *everything* within him seemed fascinating; he did not want, in the interest of mere consistency, to throw any part of himself away. In one moment he is a skeptic, in the next a dogmatic believer; a well-trained scientific mind and then a mocker of science; a man blessed with "a constitution so generall, that it consorts and sympathizeth with all things" who yet marvels at the obstinacy of the Jews, "that contemptible and degenerate issue of *Jacob*," and remarks smugly that "man is the whole world and the breath of God, woman the rib onely and crooked piece of man." Browne's insatiable curiosity is also accompanied by a deep and habitual complacency.

To be sure, Browne informs the reader that he reputes himself "the abjectest piece of mortality" and detests his own nature, and he thanks God that he has escaped at least one of the vices to which the children of Adam were prone: pride. At which point he proceeds to rehearse at some length all of the accomplishments he might boast of, were he inclined to do so: "For my owne part, besides the *Jargon* and *Patois* of severall Provinces, I understand no less then six Languages; yet I protest I have no higher conceit of my selfe than had our Fathers before the confusion of *Babel*, when there was but one Language in the world, and none to boast himselfe either Linguist or Criticke," etc.

Disingenuousness of this magnitude has its charms, as does

Browne's assertion that he "could lose an arme without a teare, and with few groans, mee thinkes, be quartered into pieces." (Mercifully, he was never put to the test.) He concluded that he was not truly known by his contemporaries: "I am in the darke to all the world, and my nearest friends behold mee but in a cloud." There was a painful discrepancy between the figure that he cut and what he felt himself to be. "Men that look upon my outside, perusing onely my condition, and fortunes," he writes, "do erre in my altitude; for I am above *Atlas* his shoulders." Such self-vaunting risked mockery, as Browne must have known, but it was his way of affirming a core feeling, an exaltation of the very fact of his existence: "Now for my life, it is a miracle of thirty yeares."

For Browne, as for Montaigne, a single life—*his* life alone—contained the whole of the human condition and indeed the whole of creation: the instinctual life of animals and the mental agility of angels, abject mortality and divine grace, exaltation and degradation, heaven and hell. "I finde there are many pieces in this one fabricke of man; and that this frame is raised upon a masse of Antipathies." It was not only the inner conflicts that were fascinating—within the compass of himself, he writes, he could find the "battell of *Lepanto*"—but also the sheer sense of multitude: "thus is man that great and true *Amphibium*, whose nature is disposed to live not onely like other creatures in divers elements, but in divided and distinguished worlds."

There is a resonant comic moment in Shakespeare's *The Tempest* in which the clown Trinculo, seeking shelter from a storm, stumbles across the inert body of the islander Caliban and wonders what he has found: "What have we here? a man or a fish? dead or alive?... Were I in England now, as once I was, and had but this fish painted, not a holiday fool there but would give a piece of silver: there would this monster make a man; any strange beast there makes a man.... Legged like a man and his fins like arms!"[8] In *Religio Medici* Browne in effect identified himself—and all men—as a marvel, a "great and true *Amphibium*" of the kind that Trinculo dreams of putting on display.

The dream of the Shakespearean clown was not far from reality: collectors in the sixteenth and seventeenth centuries went to extravagant lengths to assemble in one place curious objects from all over the globe. Some of the collections were open to the public, for a fee; some were gathered for the private gratification of the collector himself or the patron and his friends. In Oxfordshire, the Tradescants, a father and son employed as gardeners in the service of the Duke of Buckingham, gathered unusual botanical specimens, many from Africa and the New World, enough to fill "twelve cartloads of curiosities."[9] The collection seemed to contemporaries astonishingly comprehensive; it was known as "The Ark." Other great English collectors, following a fashion already well-established on the Continent, labored to create what Germans called *Wunderkammern*, wonder cabinets. The typical contents of these collections—coconut-shell goblets, elaborately carved ivory knickknacks, seashells, bits of coral, antique coins and cameos, stuffed armadillos, geodes, enormous fossilized bones, polished rocks that seemed to depict landscapes, unicorn horns, birds of paradise, aberrant fruits and monstrous animals, anamorphic pictures, mechanical ducks that quacked and flapped their wings, Indian featherwork capes, Turkish shoes, "barnacle geese" that grew on trees in Scotland, mummified hands, dragons' teeth, ostrich eggs, and so on and so forth—eventually fell out of favor as more disciplined models of scientific inquiry, organization, and proof took hold. But for Browne and like-minded contemporaries the whole point of such displays was to emphasize category confusion, to exemplify metamorphosis and occult resemblance, to arouse astonishment at the world's strangeness and infinite variety, and to collapse the distinction between art and nature.

Browne had a *Wunderkammer* of his own at his home in Norfolk. By the mid-1660s he had some forty animals on display, including curlews in cages, seafowl artificially stuffed with fish, a pelican, the heads of two whales, and boxes of flies and other insects.[10] The English diarist John Evelyn (1620–1706), who visited him there, wrote that the "whole house and garden [was] a paradise and cabinet of rarities."[11] And as a collector Browne's dreams did not stop with these objects. Among his miscellaneous papers, there is a playful catalogue for what he called a "Musaeum Clausum, or Bibliotheca Abscondita," an imaginary collection of rarities "never seen by any man now living." The catalogue fascinated Sebald, who in *The Rings of Saturn* lovingly rehearses some of its wonders:

> a drawing in chalk of the great fair of Almachara in Arabia, which is held at night to avoid the great heat of the sun; a painting of the famous battle fought between the Romans and the Jaziges on the frozen Danube; a dream image showing a prairie or sea meadow at the bottom of the Mediterranean, off the coast of Provence; Solyman the Magnificent on horseback at the siege of Vienna, and behind him a whole city of snow-white tents extending as far as the horizon; a seascape with floating icebergs upon which sit walruses, bears, foxes and a variety of rare fowls; and a number of pieces delineating the worst inhumanities in tortures for the benefit of the observer: the scaphismus of the Persians, the living truncation of the Turks, the hanging sport at the feasts of the Thracians, the exact method of flaying men alive, beginning between the shoulders, according to the meticulous description of Thomas Minadoi.[12]

The vertiginous collection of bizarre curiosities belongs to the realm of the impossible: "He who knows where all this Treasure now is," Browne concludes his list, "is a great Apollo. I'm sure I am not He."[13]

But Browne's great realization in *Religio Medici* was that he did not need to be Apollo nor did he require armies of collectors, real or imaginary. He needed only to embark on a voyage through his mind and to display in an elegant cabinet made of prose what he had found. "Wee carry with us," he writes, "the wonders, wee seeke without us: There is all *Africa*, and her prodigies in us." It was all well and good to ransack the world in search of curiosities; "the world that I regard is my selfe, it is the Microcosme of mine owne frame, that I cast mine eye on." Some at least of what seems to be Browne's absurd narcissism derives less from vanity than from literal self-regard. If man is a great amphibian, living in divided and distinguished worlds, if each individual is a universe in miniature, then Browne must necessarily discover in himself an altitude above Atlas's shoulders and a vastness greater than all the oceans of the terrestrial globe. He wants us to share his surprise at all that is within him, for that surprise may lead us to make comparable discoveries within ourselves. It is in this spirit, for example, that he asks us to marvel at the theater that he has located in his dream world: "I am no way facetious, nor disposed for the mirth and galliardize of company; yet in one dreame I can compose a whole Comedy, behold the action, apprehend the jests, and laugh my selfe awake." To join with Browne in this wonder is to begin to apprehend the magical space of our own dreams.

Many of the objects assembled in the seventeenth-century wonder cabinets became the basis of the great scientific museums that were founded at around this time and in the succeeding century. The Tradescants' cartloads of curiosities were eventually absorbed into Oxford's Ashmolean Museum, and the vast collections of Sir Hans Sloane were bequeathed to the nation as the core of the British Museum. But these scientific museums were organized, as Francis Bacon had already foreseen in *The New Atlantis* (1627), on very different principles and to very different ends from those that had inspired Browne. The rational schemes of classification and research diminished the delight in paradox, contradiction, and occult mysteries that *Religio Medici* attempts to heighten. Though he was proud of his scientific training as a doctor, Browne seems to have had little interest, apart from the circulation of the blood, in what we regard as the significant achievements of sixteenth- and seventeenth-century natural science. He seems at times to belittle them, as if they posed a vague threat he could ward off by lumping them together with a dismissive smile: "some have held that Snow is blacke, that the earth moves, that the soule is ayre, fire, water; but all this is Philosophy."

Browne in fact had a stake in mystifications of the sort that the scientific intelligence of Bacon or the theological intelligence of Calvin abhorred. He prefers a murky causality: "there is no liberty for causes to operate in a loose and stragling way," he declares, as if he believes that everything is fixed in a determinate chain, and then a moment later adds that "'Tis not a ridiculous devotion, to say a Prayer before a game at Tables." Will God intervene, in response to this prayer, and change the way the dice were going to fall, or will he not? How does Providence operate in the world? "I will not say God cannot," Browne writes, "but hee will not performe many things, which wee plainely affirme he cannot." What is being secured here through a characteristic dance of evasions is the realm of the spirit, the religious dimension that Browne set out to articulate and that he struggles throughout his work to defend.

In keeping with Browne's claim that everything is already within him, the principal challenges to which he responds are mounted by himself. "I do believe," goes a typical passage, "there was already a tree whose fruit our unhappy parents tasted, though in the same Chapter, where God forbids it, 'tis positively said, the plants of the field were not yet growne; for God had not caused it to raine upon the earth." Or again, "Experience, and History informe me, that not onely many particular women, but likewise whole Nations have escaped the curse of childbed, which God seemes to pronounce upon the whole Sex; yet doe I beleeve that all this is true, which indeed my reason would perswade me to be false." Or again, "I can read the story of the Pigeon that was sent out of the Ark, and returned no more, yet not question how shee found out her mate that was left behind: That *Lazarus* was raised from the dead, yet not demand where in the interim his soule awaited." Or again, "I would gladly know how *Moses* with an actuall fire calcin'd, or burnt the golden Calfe into powder: for that mysticall mettle of gold . . . exposed unto the violence of fire, grows onely hot and liquifies."

In all of these instances—and there are many more of them—it is Browne himself who raises the objection, calling attention to a contradiction in Scripture that others might ignore, adducing his own experience to refute various of its claims, taking the words literally and by doing so revealing their absurdity or impossibility. He knowingly voices doubts and poses problems, even when he declares that it is improper to do so: "Certainly it is not a warrantable curiosity, to examine the verity of Scripture by the concordance of humane history, or seek to confirme the Chronicle of *Hester* or *Daniel*, by the authority of *Megasthenes* or *Herodotus*." It may not be "warrantable" to call into question the historical reality of alleged miracles or observe that the Hebrew chronicles do not easily square with pagan authorities, but—warrant or no—Browne has just done so, as he himself half admits: "I confesse I have had an unhappy curiosity this way, till I laughed my selfe out of it."

"Unhappy curiosity" is not the only or the deepest hidden cause of Browne's doubts; he attributes their origin to the work of the devil. "Having seene some experiments of *Bitumen*, and having read farre more of *Naphta*, he whispered to my curiositie the fire of the Altar might be naturall; and bid me mistrust a miracle in *Elias* when he entrench'd the Altar round with water; for that inflamable substance yeelds not easily unto water, but flames in the armes of its Antagonist." Who is it in this sentence that has seen some experiments of bitumen, the Devil or the budding scientist Browne? The ambiguity is part of the point, since the father of lies is sitting at the same table with his intended victim: "the Devill playd at Chesse with mee, and yeelding a pawne, thought to gaine a Queen of me, taking advantage of my honest endeavours; and whilst I labour'd to raise the structure of my reason, hee striv'd to undermine the edifice of my faith."

Browne's skeptical intelligence and his observation of the nature of things told him relentlessly that many of the claims on which his beliefs were based were improbable or false. With his training at Oxford, Montpellier, Padua, and Leiden, he understood better than most of his contemporaries the vulnerability of the body to what Hamlet called "the thousand natural shocks that flesh is heir to." As a physician he had studied the etiology of these natural shocks. He knew that a man who one day looked perfectly healthy could the next succumb to a fatal illness; that seemingly robust children would abruptly sicken and die; that people could scream with the conviction that someone was sticking a pin in their flesh, when in fact the pain's source was entirely internal. "Men that looke no further than their outsides thinke health an appertinance unto life," he writes, "and quarrell with their constitutions for being sick." But Browne had looked not on the outsides alone, and he had concluded that there was nothing strange or uncanny about sickness or death. "I that have examined the parts of man, and know upon what tender filaments that Fabrick hangs," he exclaims, "doe wonder that we are not alwayes so; and considering the thousand dores that lead to death doe thanke my God that we can die but once."

Yet this medical doctor, so confident in his understanding of the nature of the body, insisted on crediting the wildest and most dangerous beliefs in his time about the origins of certain catastrophic illnesses, beliefs that led men and women to bring charges of witchcraft against their neighbors. By the 1640s there were many in England, in the medical establishment and elsewhere, who dismissed these charges as grotesque fantasies: people died suddenly and unexpectedly, they said, not because a witch had cast a spell upon them but because they succumbed to a hidden disease. But Browne was not one of the skeptics. "For mine owne part," he affirmed, "I have ever beleeved, and doe now know, that there are Witches." He did not, of course, have any empirical evidence. He had rather the necessity, as he understood it, of his faith. Like King James before him, he argued that those who call into question the existence of witches "doe not onely deny them, but [all] Spirits; and are obliquely and upon consequence a sort, not of Infidels, but Atheists."[14] No witches, no spirits. No spirits, no God. What did the absence of evidence count against such pressure? Indeed the demand to see evidence was itself a sign that the skeptic had already fallen under Satan's power: "Those that to confute their incredulity desire to see apparitions, shall questionlesse never behold any," Browne contends, "nor have the power to be so much as Witches; the Devill hath them already in a heresie as capitall as Witchcraft, and to appeare to them, were but to convert them." "As capitall as Witchcraft": Browne understood perfectly well the real-world consequences of his own position, which he continued to hold long after the publication of *Religio Medici.* As late as

1664 he gave evidence for the prosecution at the trial of Amy Duny and Rose Cullender, who had been charged with witchcraft.[15] Both were convicted and executed.

In *Religio Medici* Browne's mind keeps returning to the problem of unbelief, restlessly searching out the cracks in the foundations of his faith. The only way in which he can keep that faith from collapsing is to stage again and again the defeat of his doubts. That is to say—and Browne says it explicitly—the religion of the doctor depends upon the defeat of his reason. "I love to lose my selfe in a mystery," he exults, "to pursue my reason to an *o altitudo*." *"O altitudo!"*—from the Vulgate—is Saint Paul's exclamation of wonder at the profundity of God's wisdom: "how unsearchable are his judgments, and his ways past finding out" (Romans 11:33). And with that exclamation Browne's mind, so drawn to solving the "involved aenigma's and riddles" of faith, brings itself to a grinding halt: "I can answer all the objections of Satan, and my rebellious reason, with that odde resolution I learned of *Tertullian*, *Certum est quia impossibile est*." Thus the ancient Church Father Tertullian had quelled his rational doubts about the Resurrection: It is true because it is impossible.

4.

In the village of Great Walsingham in Norfolk around 1655, not far from where Browne lived, a group of men digging in a rural field unearthed between forty and fifty ancient urns full of human ashes, pieces of bones, and funerary objects. This discovery—an unusual but not unprecedented surging up of pagan remnants from Britain's seemingly solid Christian soil—was the occasion for Browne's prose meditation, *Hydriotaphia, or Urne-Buriall*, first published in 1658, and reprinted twice in Browne's lifetime (in 1659 and 1669). As a physician, Browne had been trained to "keep men out of their Urnes," or whatever other vessel might contain their bodily remains. As a physician, he also knew perhaps better than anyone how constrained that power necessarily was. Contemplating these ancient urns provoked Browne the essayist to reflect upon how different cultures handled the remains of their dead, and how they understood the posthumous fate that awaited them. The result was a piece of writing that defies easy categorization: part antiquarianism, part religious meditation, and part anthropological observation, *Urne-Buriall* is above all an extraordinary piece of writing that many readers regard as Browne's masterpiece.

What was it that attracted Browne about the discovery of ancient urns in Norfolk? At the most obvious level, the discovery appealed to his interests as an antiquary, a hobby he describes in the dedicatory epistle to *Urne-Buriall* as "run[ning] up your thoughts upon the ancient of dayes, the Antiquaries truest object, unto whom the eldest parcels are young, and earth it self an Infant." As we have seen, Browne was a great collector of ancient objects. In a letter he wrote to his son Thomas in 1661, he boasts that "my coyns are encreased since you went, I had 60 coynes of King Stephen found in a grave before Christmas, 60 Roman silver coyns I bought a month agoe, and Sir Robart Paston will send me his box of Saxon and Roman coyns next week which are about thirtie."[16] He was also captivated, like many of his contemporaries, with the layers of British history that could be discovered simply by digging a few inches into the earth. Indeed, Norfolk seemed to him a virtual treasure trove of ancient materials: coins, pots, and human ashes were all waiting to be uncovered, to breathe the English air.

Browne's interest in recovering the ancient British past was surely intensified by the instability of the contemporary world around him. *Urne-Buriall* was written toward the very end of the civil war, which in addition to costing many lives had brought great destruction to English monuments and churches. Although Browne was not a very political man, his sympathies were with the Royalists, and his instincts on the whole were deeply conservative. The urns found outside of Norfolk hearkened back for him to a nobler time in British history, when the savages were civilized by a great, foreign culture. Browne imagines the urns contained the remains either of Romans, or of "*Brittains Romanised*." In honor of the "early civility they brought upon these Countreys," he writes in the dedicatory epistle to his friend, Sir Thomas Le Gros, "We mercifully preserve their bones, and pisse not upon their ashes." It is hard not to hear in these words Browne's indictment of the Parliamentarian radicals behind much of the iconoclasm.

Browne turned out to be wrong about the provenance of the Norfolk urns: as subsequent scholarship has established, the urns were Anglo-Saxon, and probably date to around 500 CE. (With one possible exception, held in the Ashmolean Museum, none of these urns survives today.) The fact that there were no Roman coins or Latin inscriptions, combined with the modesty of the objects found (brass plates, tweezers, modest jewels), should have suggested to him that the urns were not, in fact, Roman. Browne was particularly interested, however, in connecting Britain's past with the sophisticated Romans rather than with the far less cultivated Anglo-Saxons, and this interest may have clouded his interpretive, antiquarian skills.

When Browne encountered an ancient coin or bone or urn, he treated it like a medical patient, noting every detail of its physical appearance—its size, color, and condition—as if diagnosing bodily symptoms. And in the deepest sense, *Urne-Buriall* is not simply or primarily his diagnosis of the urns at Norfolk: it is his diagnosis of the human condition. The discovery of the urns appealed to Browne's profound curiosity about death and the afterlife; it combined his fascination with anthropology with his interest in theology. At the heart of *Urne-Buriall* is a meditation on two different ways of disposing of the dead: cremation, a process that rapidly reduces the body to ashes on a funerary pyre; and inhumation, which buries the body intact, leaving it to disintegrate slowly over time. Browne associates the burning of the dead with the ancient Greeks and Romans, although he acknowledges evidence of this custom in pagan Denmark and Germany. He speculates that the urns found in Norfolk are more than 1,300 years old, dating the general end of cremation practices to sometime after the Antonine period in Rome. (Cremation was in fact only the norm in ancient Rome for a relatively short period, from around the first century BCE to the mid-second century CE, and was more typical for the lower classes, whose urns were typically placed in niches in collective burial grounds, or *columbaria*, while their social betters might still be inhumed in tombs. By the third century CE, ordinary Romans began to be buried in catacombs, which made inhumation much more affordable, and whether as a result of Christian influence or for other reasons, there was a marked shift back toward interment.)

Browne is not certain when cremation began or ended in ancient Britain; he assumes that it became a regular practice with the arrival of the Romans and ceased sometime in the early centuries after conversion to Christianity. What matters for Browne, what allows him to set up the characteristically wayward course of reflections he will pursue, is the apparently unchristian character of the funeral rites. "Christians abhorred this way of obsequies," he declares, "and though they stickt not to give their bodies to be burnt in their lives, detested that mode after death."

Browne speculates that the urns were buried "either as farewells unto all pleasure" or in "vain apprehension that they might use them in the other world." The adjective "vain" quietly but decisively dismisses the hope of bringing useful or beloved objects to the afterlife. Judgments such as this are, in fact, few and far between: one of the striking features of *Urne-Buriall* is Browne's detachment, for nearly all of the text, from any of the parties described. He adopts on the whole the distant tone of an anthropologist, speaking of pagans and Christians alike as a distanced third-person "they," as if he wanted to observe all humanity from beyond the grave. (This coolness in tone sits in an interesting tension with the text's rhetorical exuberance and grandeur.) But in this instance, Browne asserts with unusual conviction that the idea that burying objects with the dead could alter the course of the afterlife is mistaken. "That they buried a peece of money with them as a Fee of the *Elysian Ferry-man*," he writes, "was a practise full of folly."

Before assuming too pious or righteous a tone, however, Browne qualifies his criticism, donning his hat as cultural anthropologist rather than enlightened Christian. Although the habit of burying the dead with precious objects or provisions does not, in his opinion, have any metaphysical consequences, he commends this practice for providing a rich opportunity to learn more about the past. Whatever it lacked in eschatological purpose, it more than compensated for in its contribution to human knowledge. These "are laudable wayes of historicall discoveries," he concludes, "and posterity will applaud them."

Given Browne's fascination with the physical remains of the past, his main objection to cremation is that it destroys the evidence. He laments that we can determine neither the size nor the figure nor the sex of the dead once they have been reduced to ashes, nor can we read the person's character, which can be found "in sculls as well as faces." And yet the advantages of burning the body are considerable. In one of his more grossly material descriptions, Browne describes the ways in which the body, once buried, will be recycled after death: "To be gnaw'd [or knav'd] out of our graves, to have our sculs made drinking-bowls, and our bones turned into Pipes, to delight and sport our Enemies, are Tragicall abominations, escaped in burning Burials."[17] Cremation also eliminates what John Donne called, coining a new word, the "vermiculation" of the flesh: "Urnall enterrments, and burnt Reliques," Browne acknowledges, "lye not in fear of worms, or to be an heritage for Serpents."

As curious as Browne was about the identity of the people whose remains filled the urns, he was perhaps even more concerned with the religious questions the urns raised. The ancient urns were clues to pagan beliefs about the afterlife, and Browne treats these beliefs with a great degree of sympathy. Underlying *Urne-Buriall* is the tacit recognition that pagans and Christians share many of the same fears and desires when confronting their deaths, and that their respective motivations for cremation or inhumation have more in common than many of his fellow Christians might like to believe. When Browne observes, for example, that a number of the urns may contain the remains of more than one person, he connects powerfully to the longing for joint burial, however futile or vain he ultimately believed such a longing to be. In these wonderful sentences brimming with uncharacteristic emotion, he attempts to imagine why two people might choose for their ashes to be buried together:

> The ashes of *Domitian* were mingled with those of *Julia*, of *Achilles* with those of *Patroclus*: All Urnes contained not single Ashes; Without confused burnings they affectionately compounded their bones; passionately endeavouring to continue their living Unions. And when distance of death denied such conjunctions, unsatisfied affections conceived some satisfaction to be neighbours in the grave, to lye Urne by Urne, and touch but in their names.

"To lye Urne by Urne": in Browne's day as in ours, husbands and wives were routinely buried together, if not in shared tombs, then at least side by side. Hence Browne would have recognized that the desire to "continue their living Unions" through lying together in the earth was a desire that pagans and Christians shared, at the same time that he registered how irrational such a desire may be. This description gains its poignancy precisely through its recognition of futility: wanting to touch in the flesh, the dead meet only in name.

Many cultures have, of course, supposed that preserving the body and provisioning the grave somehow strengthen and equip the soul for the rigors or ordeal of the afterlife. Neither pagans nor Christians can be supposed to have imagined the dead nestling fondly together underground or in their clay pots, a prospect that Christians, at least, guarded against by teaching that once the soul departs from the flesh, the corpse has no further sentient experience.

And yet, the urge for companionship overwhelmed any such rational explanations. Browne needed hardly look back to Achilles and Patroclus or Domitian and Julia for examples of couples who wanted to be buried together; he might have turned to the work of Donne, who in his poem "The Relic" described himself and his mistress as a "loving couple" in the grave made up of a "bracelet of bright hair about the bone," and who begged in his epitaph for his wife, Anne Donne, that "his ashes to these ashes / In a new marriage (may God assent)" be joined together.

The urge for joint burial—to carry an earthly attachment across the grave—was only one manifestation of what Browne regarded as a much larger reluctance in both pagan and Christian cultures to believe in the finality of death. "It is the heaviest stone that melancholy can throw at a man," he exclaims, "to tell him he is at the end of his nature; or that there is no further state to come." It is for this reason that we wish to be remembered for what we were, a hope that is often projected onto the preservation of our physical remains. And yet, as Browne looked down upon the anonymous urns in Norfolk, which despite having survived hundreds and hundreds of years told him no more about the individuals they contained than did the dirt on the ground below him, he saw only the futility of such practices. "Vain ashes," he declares, "which in the oblivion of names, persons, times, and sexes, have found unto themselves a fruitlesse continuation, and only arise unto late posterity, as Emblemes of mortall vanities."

Browne's indictment was not limited to the pagans whose ashes he addresses directly. It applied equally if not even more so to enlightened Christians. The erection of Christian monuments to the dead is "a vanity almost out of date," for armed with knowledge of the Last Day and the subsequent Resurrection, the faithful should dispense with all concern for their earthly memory. "To extend our memories by Monuments," Browne writes, "whose death we dayly pray for, and whose duration we cannot hope, without injury to our expectations in the advent of the last day, were a contradiction to our beliefs." This is one of the rare instances in *Urne-Buriall* in which Browne identifies himself as a Christian—he refers to "*our* beliefs"—and he does so only in the context of pointing out how little sense it makes, for him as for his fellow churchmen, to pursue any means of earthly commemoration.

Like many of his generation living through the horrors of war, Browne regarded the end of the world as imminent. "'Tis too late,"

Browne exclaims, "to be ambitious." (This millenarianism interestingly crossed the political divide, affecting Parliamentarians and Royalists alike.) Browne also saw a fundamental tension between the pursuit of mortal fame and the certainty of heavenly judgment. "The sufficiency of Christian Immortality frustrates all earthly glory," he declares, "and the quality of either state after death [salvation or damnation] makes a folly of posthumous memory." Whether we end up in heaven or hell, the status of our memory on earth will be of no consequence whatever.

If only we could be assured of our happiness in the afterlife, Browne continues, we would not even bother with this world. In such circumstances, he imagines "it were a martyrdome to live." But very few people have such assurance. And even those who feel reasonably assured of their salvation, Browne insists, still experience the fear of death, which will have no bearing on their heavenly status. "The long habit of living," he comments, "indisposeth us for dying." This is a strikingly humane note to sound compared to the hard line of some of his contemporaries: English Puritans, for example, regularly preached that fearing death was a symptom of damnation, and that the manner of our death predicted the chances of our salvation. Browne, by contrast, asserts that even Christian martyrs who died in their old age may well have been fearful of death—"the constitution of old age," he reasons, "naturally makes men fearfull; complexionally superannuated from the bold and courageous thoughts of youth and fervent years"—but this complexional disadvantage has no bearing on their ultimate fate. "They may sit in the *Orchestra*, and noblest Seats of Heaven," he declares, "who have held up shaking hands in the fire, and humanly contended for glory."

If, in the end, the question of how we are buried and whether we are remembered means little or nothing, why should we bother to think about the respective advantages of cremation versus inhumation, the competing methods for preserving the corpse, and the like? What is the purpose, in other words, of disquisitions like *Urne-Buriall*? One simple answer would be that Browne's text is ultimately in the service of a kind of Christian triumphalism: whatever relativism it might sound in the middle of its pages, by the end it asserts only the superior position of Christian faith, which guarantees immortal salvation to its true believers.

But to reach this conclusion about *Urne-Buriall* is to strip it of its terrific curiosity, its willingness to examine different practices and traditions with very few prejudices or obvious preferences. It is to reduce the idiosyncratic and dynamic working of Browne's mind to the rote rehearsals of Christian truths. For however much Browne wants to affirm the certainty of Christian immortality, the effect of *Urne-Buriall* as a whole pulls in the other direction. We are left with an overwhelming sense of how difficult it is for human beings—whether Romans or Greeks or Scythians or Jews or Muslims or Christians—to accept their mortality, and how much it matters for the living, if not for the dead, that the deceased be treated with affection and respect. We are left with a sense as well that however much we might recognize rationally that our effort to secure some type of posthumous life for the flesh is futile, we cannot help but pursue it all the same.

Whether or not we will one day achieve heavenly splendor, Browne's interest finally lies in the poignancy of earthbound human nature. "Man is a noble animal," he concludes, "splendid in ashes, and pompous in the grave." It was this sentiment that De Quincey paraphrased in his ecstatic description of Browne's prose as a "melodious ascent as of a prelude to some impassioned requiem breathing from the pomps of earth, and from the sanctities of the grave!"[18] And however much *Urne-Buriall* seems to argue for the futility of human endeavors, it is also one of the great verbal testaments to how magnificent such endeavors can be. Although Browne's use of the term "pompous" has some of its modern meaning of inflated self-importance, in the seventeenth century "pompous" also had a different sense: it meant characterized by pomp, as in a stately procession. In the fifth and final chapter of *Urne-Buriall*, whose sentences achieve a lyricism unprecedented in English prose, we are met with what we might now identify as a characteristically Brownian paradox. He is calling attention to the folly of man's dreams of creating a grand, enduring posthumous memory—the "greater part" of man, he advises, "must be content to be as though they had not been, to be found in the Register of God, not in the Record of Man"—at the same time that the peculiarity and beauty and power of his own language makes a serious bid for him to be one of those remembered. In the very last paragraph, he exclaims:

> To subsist in lasting Monuments, to live in their productions, to exist in their names, and prædicament of *Chymera's*, was large satisfaction unto old expectations, and made one part of their *Elyziums*. But all this is nothing in the Metaphysicks of true belief. To live indeed is to be again our selves, which being not only an hope but an evidence in noble beleevers, 'Tis all one to lye in St *Innocents* Church-yard, as in the Sands of *Ægypt*: Ready to be any thing, in the extasie of being ever, and as content with six foot as the Moles of *Adrianus*.

Sentences like this, which move from the vain hopes of the ancients to the "Metaphysicks of true belief," from the dry sands of Egypt to the damp churchyard of St. Innocents in Paris, are entirely typical of this text in their accumulation of detail after detail before delivering the final message: we are "Ready to be any thing, in the extasie of being ever." This is the fantasy that Browne is, finally, most in touch with—the fantasy of "being ever"—and to the extent that his readers reexperience and reanimate his mind as they move through this extraordinary work, we might well conclude that his wish has been granted.

5.

Browne was fascinated by the oxymoronic yoking together of spirit and matter, sublimity and corruption, life and death. His sensibility was alert to transformations of the kind Shakespeare invokes in a famous song from *The Tempest*:

> Full fathom five, thy father lies,
> Of his bones are coral made.
> Those are pearls that were his eyes:
> Nothing of him that doth fade
> But doth suffer a sea-change
> Into something rich and strange.
> (1.2.400–405)

But Shakespeare is describing only the fate of a drowned man. Browne suggests that all humans, drowned or saved, experience just such a process of transformation. "Wee enjoy a being and life in three distinct worlds," he writes in *Religio Medici*. In the first of these, "that obscure world and wombe of our mother," we seem to exist in the "roote and soule of vegetation." Then emerging into "the scene of the world," Browne writes, "wee arise up and become another creature, performing the reasonable actions of man." Finally, we undergo another sea change, when casting off "this slough of flesh," we are "delivered into the last world" where we can fully manifest "that part of Divinity in us."

Does Browne conceive of this description as a piece of natural science or of theology? The answer is both: "Those strange and mysticall transmigrations that I have observed in Silkewormes," he writes, "turn'd my Philosophy into Divinity." The deep pleasure of Browne's prose lies in the confounding of distinct worlds, the abrupt shifting of registers, the obscure glimpse of the beyond in the here and now, the intimation of eternity in a butterfly's wing. Browne's voice is the voice of a vanished world, a world utterly routed by our own conceptions of rational inquiry, scientific proof, and common sense. But as many modern artists from Herman Melville to Jorge Luis Borges to Sebald and others have grasped, Browne, a relic of the past, is also unnervingly one of our most adventurous contemporaries. This confounding of the very old and the very new would have immensely pleased Browne himself, who cast his mind from his own time and place all the way back to Adam and mused, "the man without a Navell yet lives in me."

—STEPHEN GREENBLATT *and*
RAMIE TARGOFF

NOTES

1. Coleridge's remarks come in a letter to Sara Hutchinson (March 10, 1804), later reprinted in *Blackwood's Magazine* (November 1819). See *The Collected Works of Samuel Taylor Coleridge*, vol. 12:1, edited by George Whalley (London: Routledge & Kegan Paul, 1980), 796.

2. Several of these curious questions are considered, along with many more, in *Pseudodoxia Epidemica: or, Enquiries into Very Many Received Tenents, and Commonly Presumed Truths* which Browne first published in 1646 and then revised and expanded through five subsequent editions, the last published in 1672.

3. Sir Kenelm Digby, *Observations vpon* Religio Medici (London, 1643), 114, 115.

4. Samuel Johnson, *Christian Morals by Thomas Browne The Second Edition. With a Life of the Author, and Explanatory Notes.* (London, 1756), liii.

5. *The Collected Writings of Thomas De Quincey*, vol. 10, edited by David Masson (Edinburgh: Adam and Charles Black, 1890), 105.

6. W. G. Sebald, *The Rings of Saturn*, translated by Michael Hulse (New York: New Directions, 1998), 19.

7. *Urne-Buriall* was originally printed in 1658 together with another independent work, *The Garden of Cyrus, or the Quincunciall, Lozenge, or Net-work Plantations of the Ancients, Artificially, Naturally, Mystically Considered*. The standard scholarly edition of Browne's complete works, on which our texts of *Religio Medici and Urne-Buriall* are based, is *The Works of Thomas Browne*, edited by Geoffrey Keynes, 4 vols., (London: Faber and Faber, 1964). We gratefully acknowledge the assistance of Stephen Hequembourg who prepared the notes and the glosses for our text of Browne.

8. *The Tempest* 2.2.23–32, in *The Norton Shakespeare,* 2nd edition, edited by Stephen Greenblatt et al. (New York: W.W. Norton, 2008).

9. On the Tradescants, see *Tradescant's Rarities: Essays on the Foundation of the Ashmolean Museum, 1683*, edited by Arthur MacGregor (Oxford: Clarendon, 1983), and on the "Wonder Cabinet," see Lorraine Daston and Katharine Park, *Wonders and the Order of Nature, 1150–1750* (New York: Zone Books, 1998).

10. See E. S. Merton, "Sir Thomas Browne as Zoologist," *Osiris* 9 (1950), 413–14.

11. Quoted in Walter E. Houghton, Jr., "The English Virtuoso in the Seventeenth Century: Part II," *Journal of the History of Ideas* 3 (1942), 197.

12. Sebald, *Rings of Saturn*, 272–73. *Scaphismus* was an ancient punishment in Persia in which the victim, confined in a log or hollow tree, had his head, arms, and legs smeared with honey to attract insects.

13. Browne, edited by Keynes, 3: 119.

14. In the *Daemonologie* (2:7), King James VI wrote, "Doubtlesslie who denyeth the power of the Devill, would likewise denie the power of God, if they could for shame. For since the Devil is the very contrary opposite to God, there can be no better way to know God, than by the contrary; as by the ones power (though a creature) to admire the power of the great Creator."

15. See Victoria Silver, "'Wonders of the Invisible World': The Trial of the Lowestoft Witches," in *Sir Thomas Browne: The World Proposed*, edited by Reid Barbour and Claire Preston (New York: Oxford University Press, 2009), 118–45.

16. *The Works of Sir Thomas Browne*, 3:8

17. Geoffrey Keynes changed "knav'd" to "gnaw'd," but there are good reasons to keep the original "knav'd," especially due to its resonance with *Hamlet*, as the editor C. A. Patrides points out. In Act Five, Scene 1 of that play, Hamlet says of Yorick's skull: "How the knave jowls it to the ground" (5.1.70–71).

18. *The Collected Writings of Thomas De Quincey*, 10: 105.

à coelo salus
Religio, Medici.
Printed for Andrew Crooke. 1642. Will: Marshall sculp:

RELIGIO MEDICI

TO THE READER

CERTAINLY that man were greedy of life, who should desire to live when all the world were at an end; and he must needs be very impatient, who would repine at death in the societie of all things that suffer under it. Had not almost every man suffered by the presse; or were not the tyranny thereof become universall; I had not wanted reason for complaint: but in times wherein I have lived to behold the highest perversion of that excellent invention;* the name of his Majesty defamed, the honour of Parliament depraved, the writings of both depravedly, anticipatively, counterfeitly imprinted; complaints may seeme ridiculous in private persons, and men of my condition may be as incapable of affronts, as hopelesse of their reparations. And truly had not the duty I owe unto the importunitie of friends, and the allegeance I must ever acknowledge unto truth prevayled with me; the inactivitie of my disposition might have made these sufferings continuall, and time that brings other things to light, should have satisfied me in the remedy of its oblivion. But because things evidently false are not onely printed, but many things of truth most falsly set forth; in this latter I could not but thinke my selfe engaged: for though we have no power to redresse the former, yet in the other the reparation being within our selves, I have at present represented unto the world a full and intended copy of that Peece which was most imperfectly and surreptitiously published before.*

This I confesse about seven yeares past, with some others of affinitie thereto, for my private exercise and satisfaction, I had at leisurable houres composed; which being communicated unto one, it became common unto many, and was by transcription successively corrupted untill it arrived in a most depraved copy at the presse. He that shall peruse that worke, and shall take notice of sundry particularities and personall expressions therein, will easily discern the intention was not publick: and being a private exercise directed to my selfe, what is delivered therein was rather a memoriall unto me then an example or rule unto any other: and therefore if there bee any singularitie therein correspondent unto the private conceptions of any man, it doth not advantage them; or if dissentaneous* thereunto, it no way overthrowes them. It was penned in such a place and with such disadvantage, that (I protest) from the first setting of pen unto paper, I had not the assistance of any good booke, whereby to promote my invention or relieve my memory; and therefore there might be many reall lapses therein, which others might take notice of, and more that I suspected my selfe. It was set downe many yeares past, and was the sense of my conceptions at that time, not an immutable law unto my advancing judgement at all times, and therefore there might be many things therein plausible unto my passed apprehension, which are not agreeable unto my present selfe. There are many things delivered Rhetorically, many expressions therein meerely Tropicall* and as they best illustrate my intention; and therefore also there are many things to be taken in a soft and flexible sense, and not to be called unto the rigid test of reason. Lastly all that is contained therein is in submission unto maturer discernments, and as I have declared shall no further father them then the best and learned judgements shall authorize them; under favour of which considerations I have made its secrecie publike and committed the truth thereof to every ingenuous Reader.

—THOMAS BROWNE

THE FIRST PART

1. FOR MY Religion, though there be severall circumstances that might perswade the world I have none at all, as the generall scandall of my profession,* the naturall course of my studies, the indifferency of my behaviour, and discourse in matters of Religion, neither violently defending one, nor with the common ardour of contention opposing another; yet in despight hereof I dare, without usurpation, assume the honourable stile of a Christian: not that I meerely owe this title to the Font, my education, or the Clime wherein I was borne, as being bred up either to confirme those principles my Parents instilled into my unwary understanding; or by a generall consent proceed in the Religion of my Countrey: But that having in my riper yeares, and confirmed judgement, seene and examined all, I finde my selfe obliged by the principles of Grace, and the law of mine owne reason, to embrace no other name but this; neither doth herein my zeale so farre make me forget the generall charitie I owe unto humanity, as rather to hate then pity Turkes, Infidels, and (what is worse) the Jewes, rather contenting my selfe to enjoy that happy stile, then maligning those who refuse so glorious a title.

2. But because the name of a Christian is become too generall to expresse our faith, there being a Geography of Religions as well as of Lands, and every Clime distinguished not onely by their lawes and limits, but circumscribed by their doctrines and rules of Faith; To be particular, I am of that reformed new-cast Religion, wherein I dislike nothing but the name; of the same beliefe which our Saviour taught, the Apostles disseminated, the Fathers authorised, and the

Martyrs confirmed; but by the sinister ends of Princes, the ambition & avarice of Prelates, and the fatall corruption of times, so decaied, impaired, and fallen from its native beauty, that it required the carefull and charitable hand of these times to restore it to its primitive integrity: Now the accidentall occasion whereon, the slender meanes whereby, the low and abject condition of the person by whom so good a worke was set on foot, which in our adversaries begets contempt and scorn, fills me with wonder, and is the very same objection the insolent Pagans first cast at Christ and his Disciples.

3. Yet have I not so shaken hands with those desperate Resolutions, who had rather venture at large their decaied bottome,* then bring her in to be new trim'd in the dock; who had rather promiscuously retaine all, then abridge any, and obstinately be what they are, then what they have beene, as to stand in diameter and swords point with them: we have reformed from them, not against them; for, omitting those improperations* and termes of scurrility betwixt us, which onely difference our affections,* and not our cause, there is between us one common name and appellation, one faith, and necessary body of principles common to us both; and therefore I am not scrupulous to converse or live with them, to enter their Churches in defect of ours, and either pray with them, or for them: I could never perceive any rationall consequence from those many texts which prohibite the children of Israel to pollute themselves with the Temples of the Heathens; we being all Christians, and not divided by such detested impieties as might prophane our prayers, or the place wherein we make them; or that a resolved conscience may not adore her Creator any where, especially in places devoted to his service; where, if their devotions offend him, mine may please him; if theirs prophane it, mine may hallow it. Holy water and the Crucifix (dangerous to the common people) deceive not my judgement, nor abuse my devotion at all: I am, I confesse, naturally inclined to that, which misguided zeale termes superstition; my common conversation I do acknowledge austere, my behaviour full of rigour, sometimes not without morosity; yet at my devotion I love to use the civility of my knee, my hat, and hands, with all those outward and sensible mo-

tions, which may express or promote my invisible devotion. I should cut off my arme rather then violate a church window, then deface or demolish the memory of Saint or Martyr. At the sight of a Crosse or Crucifix I can dispence with my hat, but scarce with the thought and memory of my Saviour; I cannot laugh at, but rather pity, the fruitlesse journeys of Pilgrims, or contemne the miserable condition of Friers; for though misplaced in circumstance, there is something in it of devotion: I could never heare the *Ave Maria* Bell[1] without an elevation, or thinke it a sufficient warrant, because they erred in one circumstance, for me to erre in all: that is in silence and dumb contempt. Whilst therefore they directed their devotions to her, I offered mine to God, and rectified the errour of their prayers by rightly ordering mine owne. At a solemne Procession I have wept abundantly, while my consorts, blinde with opposition and prejudice, have fallen into an eccesse of scorne and laughter: There are questionless both in Greek, Roman, and African Churches, solemnities, and ceremonies, whereof the wiser zeales doe make a Christian use, and stand condemned by us; not as evill in themselves, but as allurements and baits of superstition to those vulgar heads that looke asquint on the face of truth, and those unstable judgements that cannot consist in the narrow point and centre of vertue without a reele or stagger to the circumference.

4. As there were many Reformers, so likewise many reformations; every Countrey proceeding in a peculiar Method according as their nationall interest together with their constitution and clime inclined them; some angrily and with extremitie, others calmely, and with mediocrity; not rending, but easily dividing the community, and leaving an honest possibility of a reconciliation; which, though peaceable Spirits doe desire, and may conceive that revolution of time, and the mercies of God may effect; yet that judgement that shall consider the present antipathies between the two extreames,

1. A Church Bell that tolls every day at 6. and 12. of the Clocke, at the hearing whereof every one in what place soever either of house or street be-takes him to his prayer, which is commonly directed to the *Virgin*.

their contrarieties in condition, affection and opinion, may with the same hopes expect an union in the poles of Heaven.

5. But to difference my self neerer,* & draw into a lesser circle: There is no Church wherein every point so squares unto my conscience, whose articles, constitutions, and customes seeme so consonant unto reason, and as it were framed to my particular devotion, as this whereof I hold my beliefe, the Church of *England*; to whose faith I am a sworne subject, and therefore in a double obligation subscribe unto her Articles, and endeavour to observe her Constitutions; no man shall reach my faith unto another Article, or command my obedience to a Canon more: whatsoever is beyond, as points indifferent,* I observe according to the rules of my private reason, or the humour and fashion of my devotion, neither believing this, because *Luther* affirmed it, nor disapproving that, because *Calvin* hath disavouched it. I condemne not all things in the Councell of *Trent*,* nor approve all in the Synod of *Dort*.* In briefe, where the Scripture is silent, the Church is my Text; where that speakes, 'tis but my Comment; where there is a joynt silence of both, I borrow not the rules of my Religion from *Rome* or *Geneva*, but the dictates of my owne reason. It is an unjust scandall of our adversaries, and a grosse error in our selves, to compute the Nativity of our Religion from *Henry* the eight, who though he rejected the Pope, refus'd not the faith of *Rome*, and effected no more then what his owne Predecessors desired and assayed in ages past, and was conceived the State of *Venice*[2] would have attempted in our dayes. It is as uncharitable a point in us to fall upon those popular scurrilities and opprobrious scoffes of the Bishop of *Rome*, to whom as a temporall Prince, we owe the duty of good language: I confesse there is cause of passion betweene us; by his sentence I stand excommunicated, Heretick is the best language he affords me; yet can no eare witnesse I ever returned to him the name of Antichrist, Man of sin, or whore of *Babylon*; It is the method of charity to suffer without reaction: those usual Satyrs, and invectives of the Pulpit may perchance produce a

2. In their quarrels with Pope Paul the fifth.

good effect on the vulgar, whose eares are opener to Rhetorick then Logick, yet doe they in no wise confirme the faith of wiser beleevers, who know that a good cause needs not to be patron'd by a passion, but can sustaine it selfe upon a temperate dispute.

6. I could never divide my selfe from any man upon the difference of an opinion, or be angry with his judgement for not agreeing with mee in that, from which perhaps within a few days I should dissent my selfe: I have no Genius to disputes in Religion, and have often thought it wisedome to decline them, especially upon a disadvantage, or when the cause of truth might suffer in the weakenesse of my patronage: where wee desire to be informed, 'tis good to contest with men above our selves; but to confirme and establish our opinions, 'tis best to argue with judgements below our own, that the frequent spoyles and victories over their reasons may settle in our selves an esteeme, and confirmed opinion of our owne. Every man is not a proper Champion for Truth, nor fit to take up the Gantlet in the cause of Veritie: Many from the ignorance of these Maximes, and an inconsiderate zeale unto Truth, have too rashly charged the troopes of error, and remaine as Trophees unto the enemies of Truth: A man may be in as just possession of Truth as of a City, and yet bee forced to surrender; tis therefore farre better to enioy her with peace, then to hazzard her on a battell: If therefore there rise any doubts in my way, I doe forget them, or at least defer them, till my better setled judgement, and more manly reason be able to resolve them; for I perceive every mans owne reason is his best *Oedipus*,* and will upon a reasonable truce, find a way to loose those bonds wherewith the subtilties of errour have enchained our more flexible and tender judgements. In Philosophy where truth seemes double-faced, there is no man more paradoxicall then my self; but in Divinity I love to keepe the road, and though not in an implicite, yet an humble faith, follow the great wheele of the Church, by which I move, not reserving any proper poles or motion from the epicycle of my owne braine; by this meanes I leave no gap for Heresies, Schismes, or Errors, of which, at present, I hope I shall not injure Truth to say, I have no taint or tincture; I must confesse my greener studies have been polluted

with two or three, not any begotten in the latter Centuries, but old and obsolete, such as could never have been revived, but by such extravagant and irregular heads as mine; for indeed Heresies perish not with their Authors, but like the River *Arethusa*,[3] though they lose their currents in one place, they rise up againe in another: one generall Councell is not able to extirpate one single Heresie, it may be canceld for the present, but revolution of time and the like aspects of Heaven, will restore it, when it will flourish till it be condemned againe; for as though there were a *Metempsuchosis*,* and the soule of one man passed into another, opinions doe finde after certaine revolutions, men and mindes like those that first begat them. To see our selves againe wee neede not looke for *Platoes* yeare,[4] every man is not onely himselfe; there have beene many *Diogenes*, and as many *Tymons*, though but few of that name; men are lived over againe; the world is now as it was in ages past; there was none then, but there hath been some one since that parallels him, and is, as it were, his revived selfe.

7. Now the first of mine was that of the Arabians, that the soules of men perished with their bodies, but should yet bee raised againe at the last day; not that I did absolutely conceive a mortality of the soule; but if that were, which faith, not Philosophy, hath yet throughly disproved, and that both entred the grave together, yet I held the same conceit thereof that wee all doe of the body, that it should rise againe. Surely it is but the merits of our unworthy natures, if wee sleepe in darkenesse, untill the last alarum: A serious reflex* upon my owne unworthinesse did make me backward from challenging this prerogative of my soule; so I might enjoy my Saviour at the last, I could with patience be nothing almost unto eternity. The second was that of *Origen*, that God would not persist in his vengeance for ever, but after a definite time of his wrath hee

3. That looseth itselfe in Greece and riseth again in Cilicie.
4. A revolution of certaine thousand yeares when all things should returne unto their former estate and he be teaching again in his schoole as when he delivered this opinion.

would release the damned soules from torture; Which error I fell into upon a serious contemplation of the great attribute of God, his mercy, and did a little cherish it in my selfe, because I found therein no malice, and a ready weight to sway me from that other extream of despaire, wherunto melancholy and contemplative natures are too easily disposed. A third there is which I did never positively maintaine or practice, but have often wished it had been consonant to Truth, and not offensive to my Religion, and that is the prayer for the dead; whereunto I was inclined from some charitable inducements, whereby I could scarce containe my prayers for a friend at the ringing out of a Bell, or behold his corpes without an oraison for his soule: 'Twas a good way me thought to be remembred by Posterity, and farre more noble then an History. These opinions I never maintained with pertinacity, or endeavoured to enveagle any mans beliefe unto mine, nor so much as ever revealed or disputed them with my dearest friends; by which meanes I neither propagated them in others, nor confirmed them in my selfe, but suffering them to flame upon their owne substance, without addition of new fuell, they went out insensibly of themselves; therefore these opinions, though condemned by lawfull Councels, were not Heresies in me, but bare Errors, and single Lapses of my understanding, without a joynt depravity of my will. Those have not only depraved understandings but diseased affections, which cannot enjoy a singularity without a Heresie, or be the author of an opinion, without they be of a Sect also; this was the villany of the first Schisme of *Lucifer*, who was not content to erre alone, but drew into his faction many Legions of Spirits; and upon this experience hee tempted only *Eve*, as well understanding the communicable nature of sin, and that to deceive but one, was tacitely and upon consequence to delude them both.

8. That Heresies should arise we have the prophecy of Christ, but that old ones should be abolished wee hold no prediction. That there must be heresies is true, not onely in our Church, but also in any other; even in Doctrines hereticall there will be super-heresies, and Arians* not onely divided from their Church, but also among themselves: for heads that are disposed unto Schisme and complexionally

propense* to innovation, are naturally indisposed for a community, nor will ever be confined unto the order or oeconomy of one body; and therefore when they separate from others they knit but loosely among themselves; nor contented with a generall breach or dichotomie with their Church, do subdivide and mince themselves almost into Atomes. 'Tis true that men of singular parts and humors have not beene free from singular opinions and conceits in all ages; retaining something not onely beside the opinion of their own Church or any other, but also any particular Author: which notwithstanding a sober judgement may doe without offence or heresie; for there are yet after all the decrees of counsells and the niceties of the Schooles, many things untouch'd, unimagin'd, wherein the libertie of an honest reason may play and expatiate with security and farre without the circle of an heresie.

9. As for those wingy mysteries in Divinity and ayery subtilties in Religion, which have unhindg'd the braines of better heads, they never stretched the *Pia Mater** of mine; me thinkes there be not impossibilities enough in Religion for an active faith; the deepest mysteries ours containes, have not only been illustrated, but maintained by syllogisme, and the rule of reason: I love to lose my selfe in a mystery, to pursue my reason to an *o altitudo.** 'Tis my solitary recreation to pose my apprehension with those involved aenigma's and riddles of the Trinity, with Incarnation and Resurrection. I can answer all the objections of Satan, and my rebellious reason, with that odde resolution I learned of *Tertullian, Certum est quia impossibile est.** I desire to exercise my faith in the difficultest points, for to credit ordinary and visible objects is not faith, but perswasion. Some beleeve the better for seeing Christ his Sepulchre, and when they have seene the Red Sea, doubt not of the miracle. Now contrarily I blesse my selfe, and am thankefull that I lived not in the dayes of miracles, that I never saw Christ nor his Disciples; I would not have beene one of the Israelites that passed the Red Sea, nor one of Christs Patients, on whom he wrought his wonders; then had my faith beene thrust upon me, nor should I enjoy that greater blessing pronounced to all that believe & saw not. 'Tis an easie and necessary beliefe to credit

what our eye and sense hath examined: I believe he was dead, and buried; and rose againe; and desire to see him in his glory, rather then to contemplate him in his Cenotaphe, or Sepulchre. Nor is this much to beleeve, as we have reason, we owe this faith unto History: they only had the happiness and advantage of a bold and noble faith, who lived before his comming, who upon obscure prophesies and mysticall Types* could raise a beliefe, and expect apparent impossibilities.

10. 'Tis true, there is an edge in all firme beliefe, and with an easie Metaphor wee may say the sword of faith; but in these obscurities I rather use it, in the adjunct the Apostle gives it, a Buckler;* under which I perceive a wary combatant may lie invulnerable. Since I was of understanding to know we know nothing my reason hath beene more pliable to the will of faith; I am now content to understand a mystery without a rigid definition in an easie and Platonick description. That allegorical description of *Hermes*,[5] pleaseth me beyond all the Metaphysicall definitions of Divines; where I cannot satisfie my reason, I love to humour my fancy; I had as lieve you tell me that *anima est angelus hominis, est Corpus Dei*, as *Entelechia; Lux est umbra Dei*, as *actus perspicui*:* where there is an obscurity too deepe for our reason, 'tis good to sit downe with a description, periphrasis, or adumbration; for by acquainting our reason how unable it is to display the visible and obvious effects of nature, it becomes more humble and submissive unto the subtilties of faith: and thus I teach my haggard* and unreclaimed reason to stoope unto the lure of faith. I do believe there was already a tree whose fruit our unhappy parents tasted, though in the same Chapter, where God forbids it, 'tis positively said, the plants of the field were not yet growne; for God had not caused it to raine upon the earth. I beleeve that the Serpent (if we shall literally understand it) from his proper form and figure, made his motion on his belly before the curse. I find the triall of the Pucellage* and Virginity of women, which God ordained the Jewes, is very fallible. Experience, and History informe me, that not onely many particular women, but likewise whole Nations have escaped

---

5. Deus est Sphæra cuius centrum ubique, circumferentia nullibi.*

the curse of childbed, which God seemes to pronounce upon the whole Sex; yet doe I beleeve that all this is true, which indeed my reason would perswade me to be false; and this I think is no vulgar part of faith to believe a thing not only above, but contrary to reason, and against the argument of our proper senses.

11. In my solitary and retired imaginations, (*Neque enim cum porticus aut me lectulus accepit, desum mihi*)* I remember I am not alone, and therefore forget not to contemplate him and his attributes who is ever with mee, especially those two mighty ones, his wisedome and eternitie; with the one I recreate, with the other I confound my understanding: for who can speake of eternitie without a solœcisme, or thinke thereof without an extasie? Time we may comprehend, 'tis but five dayes elder then our selves, and hath the same Horoscope with the world; but to retire so farre backe as to apprehend a beginning, to give such an infinite start forward, as to conceive an end in an essence that wee affirme hath neither the one nor the other; it puts my reason to Saint *Pauls* Sanctuary;* my Philosophy dares not say the Angells can doe it; God hath not made a creature that can comprehend him, 'tis the priviledge of his owne nature; *I am that I am*, was his owne definition unto *Moses*; and 'twas a short one, to confound mortalitie, that durst question God, or aske him what hee was; indeed he only is, all other things have beene or shall be, but in eternitie there is no distinction of Tenses; and therefore that terrible terme, *Predestination*, which hath troubled so many weake heads to conceive, and the wisest to explaine, is in respect to God no prescious* determination of our estates to come, but a definitive placet* of his will already fulfilled, and at the instant that he first decreed it; for to his eternitie which is indivisible, and altogether, the last Trumpe is already sounded, the reprobates in the flame, and the blessed in *Abrahams* bosome. Saint *Peter* spoke modestly, when hee said, a thousand yeares to God are but as one day: for to speake like a Philosopher, those continued instants of time which flow into a thousand yeares, make not to him one moment; what to us is to come, to his Eternitie is present, his whole duration being but one permanent point without succession, parts, flux, or division.

12. There is no Attribute that adds more difficulty to the mystery of the Trinity, where though in a relative way of Father and Son, we must deny a priority. I wonder how *Aristotle* should conceive the world eternall, or how he could make good two Eternities: his similitude of a Triangle, comprehended in a square, doth somewhat illustrate the Trinitie of our soules,* and that the Triple Unity of God; for there is in us not three, but a Trinity of soules; because there is in us, if not three distinct soules, yet different faculties, that can, and doe subsist apart in different subjects, and yet in us are so united as to make but one soule and substance; if one soule were so perfect as to informe three distinct bodies, that were a petty Trinity: conceive the distinct number of three, not divided nor separated by the intellect, but actually comprehended in its Unity, and that is a perfect Trinity. I have often admired the mysticall way of *Pythagoras*, and the secret Magicke of numbers; Beware of Philosophy, is a precept not to be received in too large a sense; for in this masse of nature there is a set of things that carry in their front,* though not in capitall letters, yet in stenography, and short Characters, something of Divinitie, which to wiser reasons serve as Luminaries in the abysse of knowledge, and to judicious beliefes, as scales and roundles* to mount the pinnacles and highest pieces of Divinity. The severe Schooles shall never laugh me out of the Philosophy of *Hermes*, that this visible world is but a picture of the invisible, wherein as in a pourtract, things are not truely, but in equivocall shapes, and as they counterfeit some more reall substance in that invisible fabrick.

13. That other attribute wherewith I recreate my devotion, is his wisedome, in which I am happy; and for the contemplation of this onely, do not repent me that I was bred in the way of study: The advantage I have of the vulgar, with the content and happinesse I conceive therein, is an ample recompence for all my endeavours, in what part of knowledg soever. Wisedome is his most beauteous attribute, no man can attaine unto it, yet *Solomon* pleased God when hee desired it. Hee is wise because hee knowes all things, and he knoweth all things because he made them all; but his greatest knowledg is in comprehending that he made not, that is himselfe. And this is also

the greatest knowledge in man. For this do I honour my own profession and embrace the counsell even of the Devill himselfe: had he read such a Lecture in Paradise as hee did at *Delphos*,[6] we had better knowne our selves, nor had we stood in feare to know him. I know he is wise in all, wonderfull in what we conceive, but far more in what we comprehend not, for we behold him but asquint upon reflex or shadow; our understanding is dimmer than *Moses* eye, we are ignorant of the backparts, or lower side of his Divinity; therefore to pry into the maze of his Counsels, is not onely folly in Man, but presumption even in Angels; there is no thread or line to guide us in that Labyrinth; like us, they are his servants, not his Senators; he holds no Councell, but that mysticall one of the Trinity, wherein, though there be three persons, there is but one minde that decrees, without contradiction; nor needs he any, his actions are not begot with deliberation, his wisedome naturally knowes what's best; his intellect stands ready fraught with the superlative and purest Idea's of goodnesse; consultation and election,* which are two motions in us, are not one in him; his actions springing from his power, at the first touch of his will. These are Contemplations Metaphysicall; my humble speculations have another Method, and are content to trace and discover those impressions hee hath left on his creatures, and the obvious effects of nature; there is no danger to profound* these mysteries, no *Sanctum sanctorum** in Philosophy: The world was made to be inhabited by beasts, but studied and contemplated by man: 'tis the debt of our reason wee owe unto God, and the homage wee pay for not being beasts; without this the world is still as though it had not been, or as it was before the sixt day when as yet there was not a creature that could conceive, or say there was a world. The wisedome of God receives small honour from those vulgar heads, that rudely stare about, and with a grosse rusticity admire his workes; those highly magnifie him whose judicious enquiry into his acts, and deliberate research of his creatures, returne the duty of a devout and learned admiration. Therefore,

6. *γνῶθισεαυτόν* nosce teipsum.*

Search while thou wilt, and let thy reason goe
To ransome truth even to the Abysse below.
Rally the scattered causes, and that line
Which nature twists be able to untwine.
It is thy Makers will, for unto none
But unto reason can he ere be knowne.
The Devills doe know thee, but those damned meteours
Build not thy glory, but confound thy creatures.
Teach my endeavours so thy workes to read,
That learning them, in thee I may proceed.
Give thou my reason that instructive flight,
Whose weary wings may on thy hands still light.
Teach me to soare aloft, yet ever so,
When neare the Sunne, to stoope againe below.
Thus shall my humble feathers safely hover,
And though neere earth, more then the heavens discover.
And then at last, when homeward I shall drive
Rich with the spoyles of nature to my hive,
There will I sit, like that industrious flye,
Buzzing thy prayses, which shall never die
Till death abrupts them, and succeeding glory
Bids me goe on in a more lasting story.

And this is almost all wherein an humble creature may endeavour to requite, and someway to retribute unto his Creator; for if not he that sayeth *Lord Lord; but he that doth the will of the Father shall be saved*; certainely our wills must bee our performances, and our intents make out our actions; otherwise our pious labours shall finde anxiety in our graves, and our best endeavours not hope, but feare a resurrection.

14. There is but one first cause, and foure second causes* of all things; some are without efficient, as God; others without matter, as Angels; some without forme, as the first matter; but every Essence, created or uncreated, hath its finall cause, and some positive end both of its Essence and operation; This is the cause I grope after in

the workes of nature, on this hangs the providence of God; to raise so beauteous a structure, as the world and the creatures thereof, was but his Art; but their sundry and divided operations with their predestinated ends, are from the treasury of his wisedome. In the causes, nature, and affections of the Eclipse of the Sunne and Moone, there is most excellent speculation; but to profound farther, and to contemplate a reason why his providence hath so disposed and ordered their motions in that vast circle, as to conjoyne and obscure each other, is a sweeter piece of reason, and a diviner point of Philosophy; therefore sometimes, and in some things there appeares to mee as much divinity in *Galen* his Books *De usu partium*, as in *Suarez** Metaphysicks: had *Aristotle* beene as curious in the enquiry of this cause* as he was of the other, hee had not left behinde him an imperfect piece of Philosophy, but an absolute tract of Divinity.

15. *Natura nihil agit frustra*,* is the onely indisputable axiome in Philosophy; there are no *Grotesques* in nature; nor any thing framed to fill up empty cantons,* and unnecessary spaces; in the most imperfect creatures, and such as were not preserved in the Arke, but having their seeds and principles in the wombe of nature, are everywhere where the power of the Sun is; in these is the wisedome of his hand discovered; Out of this ranke *Solomon* chose the object of his admiration; indeed what reason may not goe to Schoole to the wisedome of Bees, Aunts, and Spiders? what wise hand teacheth them to doe what reason cannot teach us? ruder heads stand amazed at those prodigious pieces of nature, Whales, Elephants, Dromidaries, and Camels; these I confesse, are the Colossus and Majestick pieces of her hand; but in these narrow Engines there is more curious Mathematicks, and the civilitie of these little Citizens, more neatly sets forth the wisedome of their Maker; Who admires not *Regio-Montanus** his Fly beyond his Eagle, or wonders not more at the operation of two soules in those little bodies, than but one in the trunck of a Cedar? I could never content my contemplation with those generall pieces of wonder, the flux and reflux of the sea, the encrease of Nile, the conversion of the Needle to the North; and therefore have studied to match and parallel those in the more obvious and neglected

pieces of Nature, which without further travell I can doe in the Cosmography of my selfe; wee carry with us the wonders, wee seeke without us: There is all *Africa*, and her prodigies in us; we are that bold and adventurous piece of nature, which he that studies wisely learnes in a *compendium*, what others labour at in a divided piece and endlesse volume.

16. Thus there are two bookes from whence I collect my Divinity; besides that written one of God, another of his servant Nature, that universall and publik Manuscript, that lies expans'd unto the eyes of all; those that never saw him in the one, have discovered him in the other: This was the Scripture and Theology of the Heathens; the naturall motion of the Sun made them more admire him, than its supernaturall station* did the Children of Israel; the ordinary effects of nature wrought more admiration in them, than in the other all his miracles; surely the Heathens knew better how to joyne and reade these mysticall letters, than wee Christians, who cast a more carelesse eye on these common Hieroglyphicks, and disdain to suck Divinity from the flowers of nature. Nor do I so forget God, as to adore the name of Nature; which I define not with the Schooles, the principle of motion and rest, but, that streight and regular line, that setled and constant course the wisedome of God hath ordained the actions of his creatures, according to their severall kinds. To make a revolution every day is the nature of the Sun, because it is that necessary course which God hath ordained it, from which it cannot swerve, but by a faculty from that voyce which first did give it motion. Now this course of Nature God seldome alters or perverts, but like an excellent Artist hath so contrived his worke, that with the selfe same instrument, without a new creation hee may effect his obscurest designes. Thus he sweetned the water with a wood,* preserved the creatures in the Arke, which the blast of his mouth might have as easily created: for God is like a skilfull Geometrician, who when more easily, and with one stroke of his Compasse, he might describe, or divide a right line, had yet rather doe this, though in a circle or longer way, according to the constituted and forelaid principles of his art: yet this rule of his hee doth sometimes pervert, to acquaint

the world with his prerogative, lest the arrogancy of our reason should question his power, and conclude he could not; & thus I call the effects of nature the works of God, whose hand & instrument she only is; and therefore to ascribe his actions unto her, is to devolve the honor of God, the principall agent, upon the instrument; which if with reason we may doe, then let our hammers rise up and boast they have built our houses, and our pens receive the honour of our writings. I hold there is a general beauty in all the works of God, and therefore no deformity in any kind or species of creature whatsoever: I cannot tell by what Logick we call a Toad, a Beare, or an Elephant, ugly; they being created in those outward shapes and figures which best expresse the actions of their inward formes; and having past with approbation that generall visitation of God, who saw that all that he had made was good, that is, conformable to his will, which abhors deformity, and is the rule of order and beauty. There is therefore no deformity but in monstrosity, wherein notwithstanding there is a kind of beauty, Nature so ingeniously contriving those irregular parts, as they become sometimes more remarkable than the principall Fabrick. To speake yet more narrowly, there was never anything ugly, or mis-shapen, but the Chaos; wherein not withstanding, to speake strictly, there was no deformity, because no forme; nor was it yet impregnate by the voyce of God: Now nature is not at variance with art, nor art with nature; they being both the servants of his providence: Art is the perfection of Nature: Were the world now as it was the sixt day, there were yet a Chaos: Nature hath made one world, and Art another. In briefe, all things are artificiall, for Nature is the Art of God.

17. This is the ordinary and open way of his providence, which art and industry have in a good part discovered, whose effects wee may foretell without an Oracle; To foreshew these is not Prophesie, but Prognostication. There is another way full of Meanders and Labyrinths, whereof the Devill and Spirits have no exact Ephemerides,* and that is a more particular and obscure method of his providence, directing the operations of individualls and single Essences; this we call Fortune, that serpentine and crooked line, whereby he drawes

those actions that his wisedome intends in a more unknowne and secret way. This cryptick and involved method of his providence have I ever admired, nor can I relate the history of my life, the occurrences of my dayes, the escapes of dangers, and hits of chance with a *Bezo las Manos** to Fortune, or a bare Gramercy* to my good starres: *Abraham* might have thought that the Ram in the thicket* came thither by accident; humane reason would have said that meere chance conveyed *Moses* in the Arke* to the sight of *Pharaohs* daughter; what a Labyrinth is there in the story of *Joseph*,* able to convert a Stoick? Surely there are in every mans life certaine rubs, doublings and wrenches* which pass a while under the effects of chance, but at the last, well examined, prove the meere hand of God: 'Twas not dumbe chance, that to discover the Fougade or Powder Plot,* contrived the letter. I like the victory of 88. the better for that one occurrence which our enemies imputed to our dishonour, and the partiality of Fortune: to wit, the tempests, and contrarietie of winds. King *Philip* did not detract from the Nation, when he said, he sent his Armado to fight with men, and not to combate with the winds. Where there is a manifest disproportion between the powers and forces of two severall agents, upon a maxime of reason wee may promise the victory to the superiour; but when unexpected accidents slip in, and unthought of occurrences intervene, these must proceed from a power that owes no obedience to those axioms: where, as in the writing upon the wall, we behold the hand, but see not the spring that moves it. The successe of that pety Province of Holland (of which the Grand Seignieur proudly said, That if they should trouble him as they did the Spaniard, hee would send his men with shovels and pick-axes and throw it into the Sea) I cannot altogether ascribe to the ingenuity and industry of the people, but to the mercy of God, that hath disposed them to such a thriving *Genius*; and to the will of his providence, that dispenseth her favour to each Countrey in their preordinate season. All cannot be happy at once; for, because the glory of one State depends upon the ruine of another, there is a revolution and vicissitude of their greatnesse, which much obey the swing of that wheele, not moved by intelligences,* but by the hand of God, whereby all States arise of their Zenith and verticall points, according to their predestinated periods. For the lives not onely of men, but of Commonweales, and the whole world, run not upon a Helix that still enlargeth;* but on a Circle, where, arriving to their Meridian, they decline in obscurity, and fall under the Horizon againe.

18. These must not therefore bee named the effects of fortune, but in a relative way, and as we terme the workes of nature. It was the ignorance of mans reason that begat this very name, and by a carelesse terme miscalled the providence of God: for there is no liberty for causes to operate in a loose and stragling way, nor any effect whatsoever, but hath its warrant from some universall or superiour cause. 'Tis not a ridiculous devotion, to say a Prayer before a game at Tables; for even in *sortilegies** and matters of greatest uncertainty, there is a setled and preordered course of effects; 'tis we that are blind, not fortune: because our eye is too dim to discover the mystery of her effects, we foolishly paint her blind, & hoodwink the providence of the Almighty. I cannot justifie that contemptible Proverb, *That fooles onely are fortunate*; or that insolent Paradox, *That a wise man is out of the reach of fortune;* much lesse those opprobrious Epithets of Poets, *Whore*, *Baud*, and *Strumpet*: 'Tis, I confesse, the common fate of men of singular gifts of mind, to be destitute of those of fortune; which doth not any way deject the spirits of wiser judgements, who throughly understand the justice of this proceeding; and being enriched with higher donatives,* cast a more carelesse eye on these vulgar parts of felicity. 'Tis a most unjust ambition, to desire to engrosse* the mercies of the Almighty, not to be content with the goods of the mind, without a possession of those of the body or fortune: and 'tis an errour worse than heresie, to adore these complementall & circumstantiall pieces of felicity, and undervalue those perfections and essentiall points of happinesse, wherein we resemble our Maker. To wiser desires 'tis satisfaction enough to deserve, though not to enjoy, the favours of fortune; let providence provide for fooles: 'tis not partiality, but equity in God, who deales

with us but as our naturall parents; those that are able of body and mind, he leaves to their deserts; to those of weaker merits hee imparts a larger portion, and pieces out the defect of the one by the excesse of the other. Thus have wee no just quarrell with Nature, for leaving us naked, or to envie the hornes, hoofs, skins, and furs of other creatures, being provided with reason, that can supply them all. Wee need not labour with so many arguments to confute judiciall Astrology;* for if there be a truth therein, it doth not injure Divinity; if to be born under *Mercury* disposeth us to be witty, under *Jupiter* to be wealthy, I doe not owe a knee unto these, but unto that mercifull hand that hath disposed and ordered my indifferent and uncertaine nativity unto such benevolous aspects. Those that held that all things were governed by fortune had not erred, had they not persisted there: The Romans that erected a Temple to Fortune, acknowledged therein, though in a blinder way, somewhat of Divinity; for, in a wise supputation,* all things begin and end in the Almighty. There is a neerer way to heaven than *Homers* chaine;* an easie Logick may conjoyne heaven and earth in one argument, and with lesse than a Sorites* resolve all things into God. For though wee Christen effects by their most sensible and nearest causes, yet is God the true and infallible cause of all, whose concourse, though it be generall, yet doth it subdivide it selfe into the particular actions of everything, and is that spirit, by which each singular essence not onely subsists, but performes its operations.

19. The bad construction and perverse comment on these paire of second causes, or visible hands of God, have perverted the devotion of many unto Atheisme; who forgetting the honest advisoes* of faith, have listened unto the conspiracie of Passion and Reason. I have therefore alwayes endeavoured to compose those fewds and angry dissentions between affection, faith, and reason: For there is in our soule a kind of Triumvirate, or Triple government of three competitors, which distract the peace of this our Common-wealth, not lesse than did that other the State of Rome.

As Reason is a rebell unto Faith, so Passion unto Reason: As the propositions of Faith seeme absurd unto Reason, so the Theorems of Reason unto passion, and both unto Faith; yet a moderate and peaceable discretion may so state and order the matter, that they may bee all Kings, and yet make but one Monarchy, every one exercising his Soveraignty and Prerogative in a due time and place, according to the restraint and limit of circumstance. There are, as in Philosophy, so in Divinity, sturdy doubts, and boysterous objections, wherewith the unhappinesse of our knowledge too neerely acquainteth us. More of these no man hath knowne than my selfe, which I confesse I conquered, not in a martiall posture, but on my knees. For our endeavours are not onely to combate with doubts, but alwayes to dispute with the Devill; the villany of that spirit takes a hint of infidelity from our Studies, and by demonstrating a naturality in one way, makes us mistrust a miracle in another. Thus having perus'd the Archidoxis* and read the secret Sympathies of things, he would disswade my beliefe from the miracle of the Brazen Serpent,* make me conceit that image work'd by Sympathie, and was but an Aegyptian tricke to cure their diseases without a miracle. Againe, having seene some experiments of *Bitumen*, and having read farre more of *Naphta*, he whispered to my curiositie the fire of the Altar might be naturall; and bid me mistrust a miracle in *Elias** when he entrench'd the Altar round with water; for that inflamable substance yeelds not easily unto water, but flames in the armes of its Antagonist: and thus would hee inveagle my beliefe to thinke the combustion of *Sodom* might be naturall, and that there was an Asphaltick and Bituminous nature in that Lake before the fire of *Gomorrha*: I know that Manna is now plentifully gathered in *Calabria*, and *Josephus* tels me, in his dayes 'twas as plentifull in *Arabia*; the Devill therefore made the *quere*, Where was then the miracle in the dayes of *Moses?* the Israelites saw but that in his time, the natives of those Countries behold in ours. Thus the Devill playd at Chesse with mee, and yeelding a pawne, thought to gaine a Queen of me, taking advantage of my honest endeavours; and whilst I labour'd to raise the structure of my reason, hee striv'd to undermine the edifice of my faith.

20. Neither had these or any other ever such advantage of me, as to incline me to any point of infidelity or desperate positions of Atheisme; for I have beene these many yeares of opinion there was never any. Those that held Religion was the difference of man from beasts, have spoken probably, and proceed upon a principle as inductive as the other: That doctrine of *Epicurus*, that denied the providence of God, was no Atheism, but a magnificent and high-strained conceit of his Majesty, which hee deemed too sublime to minde the triviall actions of those inferiour creatures: That fatall necessitie of the Stoickes, is nothing but the immutable Law of his will. Those that heretofore denied the Divinitie of the holy Ghost, have beene condemned but as Heretickes; and those that now deny our Saviour (though more than Hereticks) are not so much as Atheists: for though they deny two persons in the Trinity, they hold as we do, that there is but one God.

That villain and Secretary of Hell, that composed that miscreant piece of the three Impostors;[7] though divided from all Religions, and was neither Jew, Turk, nor Christian, was not a positive Atheist. I confesse every Countrey hath its *Machiavell*, every age its *Lucian*, whereof common heads must not heare, nor more advanced judgements too rashly venture on: 'tis the Rhetorick of Satan, and may pervert a loose or prejudicate beleefe.*

21. I confesse I have perused them all, and can discover nothing that may startle a discreet beliefe: yet are there heads carried off with the wind and breath of such motives. I remember a Doctor in Physick of Italy, who could not perfectly believe the immortality of the soule, because *Galen* seemed to make a doubt thereof. With another I was familiarly acquainted in France, a Divine and a man of singular parts, that on the same point was so plunged and gravelled* with three lines of Seneca,[8] that all our Antidotes, drawne from both Scripture and Philosophy, could not expell the poyson of his errour.

---

7. Moses, Christ, and Mahomet.*

8. Post mortem nihil est, ipsaque mors nihil. Mors individua est noxia corpori, Nec patiens animæ—Toti morimur, nullaque pars manet Nostri—. Troad 399, etc.*

There are a set of heads, that can credit the relations of Mariners, yet question the testimony of Saint *Paul*; and peremptorily maintaine the traditions of *Ælian* or *Pliny*, yet in Histories of Scripture, raise Quere's and objections, beleeving no more than they can parallel in humane Authors. I confesse there are in Scripture stories that doe exceed the fables of Poets, and to a captious Reader sound like *Garagantua* or *Bevis*:* Search all the Legends of times past, and the fabulous conceits of these present, and 'twill bee hard to find one that deserves to carry the buckler unto *Sampson*; yet is all this of an easie possibility, if we conceive a divine concourse or an influence but from the little finger of the Almighty. It is impossible that, either in the discourse of man, or in the infallible voyce of God, to the weakenesse of our apprehension, there should not appeare irregularities, contradictions, and antinomies: my selfe could shew a catalogue of doubts, never yet imagined nor questioned by any, as I know, which are not resolved at the first hearing; not queries fantastick, or objections of ayre: For I cannot heare of Atoms in Divinity. I can read the story of the Pigeon that was sent out of the Ark, and returned no more, yet not question how shee found out her mate that was left behind: That *Lazarus* was raised from the dead, yet not demand where in the interim his soule awaited; or raise a Law-case, whether his heire might lawfully detaine his inheritance, bequeathed unto him by his death; and he, though restored to life, have no Plea or title unto his former possessions. Whether *Eve* was framed out of the left side of *Adam*, I dispute not, because I stand not yet assured which is the right side of a man, or whether there be any such distinction in Nature; that she was edified* out of the ribbe of *Adam* I believe, yet raise no question who shall arise with that ribbe at the Resurrection; whether *Adam* was an Hermaphrodite as the Rabbines contend upon the letter of the Text, because it is contrary to all reason, that there should bee an Hermaphrodite before there was a woman, or a composition of two natures, before there was a second composed. Likewise, whether the world was created in Autumne, Summer, or Spring, because it was created in them all; for whatsoever Signe the Sun possesseth, those foure seasons are actually existent: It is the nature of this Luminary to distinguish the severall seasons of the yeare, all which it makes at one time in the whole earth, and successively in any part thereof. There are a bundle of curiosities, not onely in Philosophy, but in Divinity, proposed and discussed by men of most supposed abilities, which indeed are not worthy of our vacant houres, much lesse our more serious studies; Pieces onely fit to be placed in *Pantagruels** Library,[9] or bound up with *Tartaretus De modo Cacandi*.

22. These are niceties that become not those that peruse so serious a Mystery. There are others more generally questioned and called to the Barre, yet me thinkes of an easie, and possible truth. 'Tis ridiculous to put off, or drowne the generall Flood of *Noah* in that particular inundation of *Deucalion*:* that there was a Deluge once, seemes not to mee so great a miracle, as that there is not one alwayes. How all the kinds of Creatures, not only in their owne bulks, but with a competency of food & sustenance, might be preserved in one Arke, and within the extent of three hundred cubits, to a reason that rightly examines it, will appeare very forcible. There is another secret, not contained in the Scripture, which is more hard to comprehend, & put the honest Father to the refuge of a Miracle; and that is, not onely how the distinct pieces of the world, and divided Ilands should bee first planted by men, but inhabited by Tygers, Panthers and Beares. How *America* abounded with beasts of prey, and noxious Animals, yet contained not in it that necessary creature, a Horse. By what passage those, not onely Birds, but dangerous and unwelcome Beasts came over: How there bee creatures there, which are not found in this triple Continent; all which must needs bee stranger unto us, that hold but one Arke, and that the creatures began their progresse from the mountaines of *Ararat*. They who, to salve this, would make the Deluge particular,* proceed upon a principle that I can no way grant; not onely upon the negative of holy Scriptures, but of mine owne reason, whereby I can make it probable, that the world was as well peopled in the time of *Noah* as in ours, and fifteene hundred

9. In Rabelais the French author.

yeares to people the world, as full a time for them as foure thousand since have beene to us. There are other assertions and common tenents drawn from Scripture, and generally beleeved as Scripture; whereunto, notwithstanding, I would never betray the libertie of my reason. 'Tis a postulate to me, that *Methusalem* was the longest liv'd of all the children of *Adam*, and no man will bee able to prove it; when from the processe of the Text I can manifest it may be otherwise. That *Judas* perished by hanging himself, there is no certainety in Scripture, though in one place[10] it seemes to affirme it, and by a doubtfull word hath given occasion so to translate it; yet in another place, in a more punctuall description, it makes it improbable, and seemes to overthrow it. That our Fathers, after the Flood, erected the Tower of *Babell*, to preserve themselves against a second Deluge, is generally opinioned and beleeved; yet is there another intention of theirs expressed in Scripture: Besides, it is improbable from the circumstance of the place, that is, a plaine in the land of *Shinar*. These are no points of Faith, and therefore may admit a free dispute. There are yet others, and those familiarly concluded from the Text, wherein (under favour) I see no consequence. To instance in one or two: as, to prove the Trinity from the speech of God, in the plural number,—faciamus hominem, Let us make man, which is but the common style of Princes, and men of Eminency;—he that shall read one of his Majesty's Proclamations may with the same logick conclude, there be two kings in England. To inferre the obedient respect of wives to their husbands from the example of Sarah, who usually called her husband Lord, which if you examine you shall finde to be no more than Seigneur or Monsieur, which are the ordinary language all civill nations use in their familiar compellations,* not to their superiours or equalls, but to their inferiours also and persons of lower condition. The Church of Rome confidently proves the opinion of Tutelary Angels,* from that answer when Peter knockt at the doore, '*Tis not he but his Angel*; that is, might some say, his Messenger, or some body from him; for so the Originall signifies, and is

10. Matt. 27.5

as likely to be the doubtfull Families meaning. This exposition I once suggested to a young Divine, that answered upon this point, to which I remember the *Franciscan* Opponent replyed no more, but, That it was a new and no authentick interpretation.

23. These are but the conclusions, and fallible discourses of man upon the word of God, for such I doe verily beleeve the holy Scriptures; yet were it of man, I could not choose but say, it is one of the most singular, and superlative Pieces that hath been extant since the Creation; were I a Pagan I should not refraine the Lecture of it; and cannot but commend the judgement of *Ptolomy*, that could not think his Library compleate without it: the Alcoran of the Turks (I speake without prejudice) is an ill composed Piece, containing in it vaine and ridiculous errours in Philosophy, impossibilities, fictions, and vanities beyond laughter maintained by evident and open Sophismes, the policy of Ignorance, deposition of Universities, and banishment of Learning, that hath gotten foot by armes and violence; This without a blow* hath disseminated it selfe through the whole earth. It is not unremarkable what *Philo* first observed, That the law of *Moses* continued two thousand yeares without the least alteration; whereas, we see, the Laws of other Common-weales to alter with occasions; and even those that pretended their originall from some Divinity, to have vanished without trace or memory. I beleeve, besides *Zoroaster*, there were divers that writ before *Moses*, who notwithstanding have suffered the common fate of time. Mens Workes have an age like themselves; and though they out-live their Authors, yet have they a stint and period to their duration: This onely is a Worke too hard for the teeth of time, and cannot perish but in those generall flames, when all things shall confesse their ashes.

24. I have heard some with deepe sighs lament the lost lines of *Cicero*; others with as many groanes deplore the combustion of the Library of *Alexandria*; for my owne part, I thinke there be too many in the world, and could with patience behold the urne and ashes of the *Vatican*, could I with a few others recover the perished leaves of *Solomon*, the sayings of the Seers, and the Chronicles of the Kings of Judas. I would not omit a Copy of *Enochs* Pillars,* had they many

neerer Authors than *Josephus*, or did not relish somewhat of the Fable. Some men have written more than others have spoken; *Pineda*[11] quotes more Authors in one worke, than are necessary in a whole world. Of those three great inventions of *Germany*,* there are two which are not without their incommodities, and 'tis disputable whether they exceed not their use and commodities. 'Tis not a melancholy *Utinam** of mine owne, but the desire of better heads, that there were a generall Synod; not to unite the incompatible differences of Religion, but for the benefit of learning, to reduce it as it lay at first in a few and solid Authours; and to condemne to the fire those swarms and millions of *Rhapsodies*, begotten onely to distract and abuse the weaker judgements of Scholars, and to maintaine the Trade and Mystery of Typographers.

25. I cannot but wonder with what exceptions the *Samaritanes* could confine their beliefe to the *Pentateuch*, or five Books of *Moses*. I am amazed at the Rabbinicall Interpretations of the Jews, upon the Old Testament, as much as their defection from the New: and truely it is beyond wonder, how that contemptible and degenerate issue of *Jacob*, once so devoted to Ethnick* Superstition, and so easily seduced to the Idolatry of their Neighbours, should now in such an obstinate and peremptory beliefe, adhere unto their owne Doctrine, expect impossibilities, and in the face and eye of the Church persist without the least hope of conversion: This is a vice in them, but were a virtue in us; for obstinacy in a bad cause, is but constancy in a good. And herein I must accuse those of my own Religion; for there is not any of such a fugitive faith, such an unstable belief, as a Christian; none that do so oft transforme themselves, not into severall shapes of Christianity and of the same Species, but into more unnaturall and contrary formes of Jew and Mahometan; that from the name of Saviour can descend to the bare terme of Prophet; and from an old beliefe that he is come, fall to a new expectation of his comming: It is the promise of Christ to make us all one flock; but how

---

11. *Pineda* in his *Monarchia Ecclesiastica* quotes one thousand and fortie Authors.
Gunnes, Printing, the Mariner's Compass.

and when this union shall be, is as obscure to me as the last day. Of those foure members of Religion wee hold a slender proportion; there are I confesse some new additions, yet small to those which accrew to our Adversaries, and those onely drawne from the revolt of Pagans, men but of negative impieties, and such as deny Christ, but because they never heard of him: But the Religion of the Jew is expresly against the Christian, and the Mahometan against both; for the Turke, in the bulke hee now stands is beyond all hope of conversion; if hee fall asunder there may be conceived hopes, yet not without strong improbabilities. The Jew is obstinate in all fortunes; the persecutions of fifteene hundred yeares have but confirmed them in their errour: they have already endured whatsoever may be inflicted, and have suffered, in a bad cause, even to the condemnation of their enemies. Persecution is a bad and indirect way to plant Religion; It hath beene the unhappy method of angry devotions, not onely to confirme honest Religion, but wicked Heresies, and extravagant opinions. It was the first stone and basis of our Faith, none can more justly boast of persecutions, and glory in the number and valour of Martyrs; For, to speake properly, those are true and almost onely examples of fortitude: Those that are fetch'd from the field, or drawne from the actions of the Campe, are not oft-times so truely precedents of valour as audacity, and at the best attaine but to some bastard piece of fortitude: If wee shall strictly examine the circumstances and requisites which *Aristotle* requires to true and perfect valour, we shall finde the name onely in his Master *Alexander*, and as little in that Roman Worthy, *Julius Cæsar*; and if any, in that easie and active part, have done so nobly as to deserve that name, yet in the passive and more terrible piece these have surpassed, and in a more heroicall way may claime the honour of that Title. 'Tis not in the power of every honest faith to proceed thus farre, or passe to Heaven through the flames; every one hath it not in that full measure, nor in so audacious and resolute a temper, as to endure those terrible tests and trialls, who notwithstanding in a peaceable way doe truely adore their Saviour, and have (no doubt) a faith acceptable in the eyes of God.

26. Now as all that die in warre are not termed Souldiers, so neither can I properly terme all those that suffer in matters of Religion Martyrs. The Councell of *Constance** condemnes *John Husse* for an Heretick; the Stories of his owne party stile him a Martyr; He must needs offend the Divinity of both, that sayes hee was neither the one nor the other: There are questionlesse many canonized on earth, that shall never be called Saints in Heaven; and have their names in Histories and Martyrologies, who in the eyes of God, are not so perfect Martyrs as was that wise Heathen,[12] that suffered on a fundamentall point of Religion, the Unity of God. I have often pitied the miserable Bishop[13] that suffered in the cause of *Antipodes*, yet cannot choose but accuse him of as much madnesse, for exposing his living on such a trifle, as those of ignorance and folly that condemned him. I think my conscience will not give me the lie, if I say, there are not many extant that in a noble way feare the face of death lesse than my selfe; yet, from the morall duty I owe to the Commandements of God, and the naturall respect that I tender unto the conservation of my essence and being, I would not perish upon a Ceremony, Politick point or indifferency: nor is my beleefe of that untractable temper, as not to bow at their obstacles, or connive at matters wherein there are not manifest impieties: The leaven therefore and ferment of all, not onely Civill, but Religious actions, is wisedome; without which, to commit our selves to the flames is Homicide, and (I feare) but to passe through one fire into another.

27. That Miracles are ceased, I can neither prove, nor absolutely deny, much lesse define the time and period of their cessation; that they survived Christ, is manifest upon record of Scripture; that they out-lived the Apostles also, and were revived at the conversion of Nations, many yeares after, we cannot deny, if wee shall not question those Writers whose testimonies wee doe not controvert, in points that make for our owne opinions; therefore that may have some truth in it that is reported of the Jesuites and their Miracles in the

12. Socrates
13. Virgilius*

Indies, I could wish it were true, or had any other testimony then their owne Pennes: they may easily beleeve those Miracles abroad, who daily conceive a far greater at home; the transmutation of those visible elements into the body and blood of our Saviour: for the conversion of water into wine, which he wrought in *Cana*, or what the Devill would have had him do in the wildernesse, of stones into Bread, compared to this, will scarce deserve the name of a Miracle: Though indeed, to speake strictly, there is not one Miracle greater than another, they being the extraordinary effects of the hand of God, to which all things are of an equall facility; and to create the world as easie as one single creature. For this is also a miracle, not onely to produce effects against, or above Nature, but before Nature; and to create Nature as great a miracle, as to contradict or transcend her. Wee doe too narrowly define the power of God, restraining it to our capacities. I hold that God can doe all things, how he should work contradictions I do not understand, yet dare not therefore deny. I cannot see why the Angel of God should question *Esdras** to recall the time past, if it were beyond his owne power; or that God should pose mortalitie in that, which hee could not performe himselfe. I will not say God cannot, but hee will not performe many things, which wee plainely affirme he cannot: this I am sure is the mannerliest proposition, wherein notwithstanding I hold no Paradox. For strictly his power is but the same with his will, and they both with all the rest doe make but one God.

28. Therefore that Miracles have beene I doe beleeve; that they may yet bee wrought by the living I doe not deny: but have no confidence in those which are fathered on the dead; and this hath ever made me suspect the efficacy of reliques,* examine the bones, question the habits and appertinencies of Saints and even of Christ himselfe: I cannot conceive why the Crosse that *Helena* found and whereon Christ himself died should have power to restore others unto life; I excuse not *Constantine* from a fall off his horse, or a mischiefe from his enemies, upon the wearing those nayles on his bridle which our Saviour bore upon the Crosse in his hands: I compute among your *Piae fraudes*,* nor many degrees before consecrated

swords and roses, that which *Baldwin* King of Jerusalem return'd the *Genovese* for their cost and paines in his warre, to wit the ashes of *John* the Baptist. Those that hold the sanctitie of the soules doth leave behind a tincture and sacred facultie on their bodies, speake naturally of Miracles, and doe not salve the doubt. Now one reason I tender so little devotion unto reliques is, I think, the slender and doubtfull respect I have alwayes held unto Antiquities: for that indeed which I admire is farre before antiquity, that is Eternity, and that is God himselfe; who though hee be stiled the Antient of dayes, cannot receive the adjunct of antiquity, who was before the world, and shall be after it, yet is not older then it: for in his yeares there is no Climacter,* his duration is eternity, and farre more venerable then antiquitie.

29. But above all the rest, I wonder how the curiositie of wiser heads could passe that great and indisputable miracle of the cessation of Oracles:* and in what swoun their reasons lay, to content themselves and sit downe with such far-fetch'd and ridiculous reasons as *Plutarch* alleadgeth for it. The Jewes that can beleeve the supernaturall solstice of the Sunne in the dayes of *Joshua*, have yet the impudence to deny the Eclipse, which even Pagans confessed at his death: but for this, it is evident beyond all contradiction,[14] the Devill himselfe confessed it. Certainly it is not a warrantable curiosity, to examine the verity of Scripture by the concordance of humane history, or seek to confirme the Chronicle of *Hester** or *Daniel*, by the authority of *Megasthenes* or *Herodotus*. I confesse I have had an unhappy curiosity this way, till I laughed my selfe out of it with a piece of *Justine*, where hee delivers that the children of *Israel* for being scabbed were banished out of Egypt. And truely since I have understood the occurrences of the world, and know in what counterfeit shapes & deceitfull vizzards times present represent on the stage things past; I doe beleeve them little more than things to come. Some have beene of my opinion, and endeavoured to write the History of their own lives; wherein *Moses* hath outgone them all, and

14. In his Oracle of *Augustus*.

left not onely the story of his life, but as some will have it of his death also.

30. It is a riddle to me, how this very story of Oracles hath not worm'd out of the world that doubtfull conceit of Spirits and Witches; how so many learned heads should so farre forget their Metaphysicks, and destroy the Ladder and scale of creatures, as to question the existence of Spirits: for my owne part, I have ever beleeved, and doe now know, that there are Witches; they that doubt of these, doe not onely deny them, but Spirits; and are obliquely and upon consequence a sort, not of Infidels, but Atheists. Those that to confute their incredulity desire to see apparitions, shall questionlesse never behold any, nor have the power to be so much as Witches; the Devill hath them already in a heresie as capitall as Witchcraft, and to appeare to them, were but to convert them: Of all the delusions wherewith he deceives mortalitie, there is not any that puzleth mee more than the Legerdemain of *Changelings*; I doe not credit those transformations of reasonable creatures into beasts, or that the Devill hath the power to transpeciate a man into a horse, who tempted Christ (as a triall of his Divinitie) to convert but stones into bread. I could beleeve that Spirits use with man the act of carnality, and that in both sexes; I conceive they may assume, steale, or contrive a body, wherein there may be action enough to content decrepit lusts or passion to satisfie more active veneries; yet in both, without a possibility of generation: and therefore that opinion, that Antichrist should be borne of the Tribe of *Dan* by conjunction with the Devill, is ridiculous, and a conceit fitter for a Rabbin than a Christian. I hold that the Devill doth really possesse some men, the spirit of melancholy others, the spirit of delusion others; that as the Devill is concealed and denyed by some, so God and good Angels are pretended by others, whereof the late detection* of the Maid of Germany[15] hath left a pregnant example.

31. Againe, I beleeve that all that use sorceries, incantations, and spells, are not Witches, or as we terme them, Magicians; I conceive

15. That lived without meat upon the smell of a Rose.*

there is a traditionall Magicke, not learned immediately from the Devill, but at second hand from his Schollers; who having once the secret betrayed, are able, and doe emperically practice without his advice, they both proceeding upon the principles of nature: where actives aptly conjoyned to disposed passives, will under any Master produce their effects. Thus I thinke a great part of Philosophy was at first Witchcraft; which being afterward derived from one to another, proved but Philosophy, and was indeed no more than the honest effects of Nature: What invented by us is Philosophy, learned from him is Magicke. Wee doe surely owe the honour of the discovery of many secrets both to good and bad Angels. I could never passe that sentence of *Paracelsus* without an asterisk or annotation; *Ascendens constellatum multa revelat, quærentibus magnalia naturæ*,*[16] i.e. *opera Dei.* I doe thinke that many mysteries ascribed to our owne inventions, have beene the courteous revelations of Spirits; for those noble essences in heaven beare a friendly regard unto their fellow-natures on earth; and therefore I beleeve that those many prodigies and ominous prognostickes which forerun the ruine of States, Princes, and private persons, are the charitable premonitions of good Angels, which more carelesse enquiries terme but the effects of chance and nature.

32. Now, besides these particular and divided Spirits, there may be (for ought I know) an universall and common Spirit to the whole world. It was the opinion of *Plato*, and it is yet of the *Hermeticall* Philosophers;* if there be a common nature that unites and tyes the scattered and divided individuals into one species, why may there not bee one that unites them all? However, I am sure there is a common Spirit that playes within us, yet makes no part of us, and that is the Spirit of God, the fire and scintillation of that noble and mighty Essence, which is the life and radicall heat of spirits, and those essences that know not the vertue of the Sunne; a fire quite contrary to the fire of Hell: This is that gentle heate that brooded on the waters,[17]

16. Thereby is meant our good Angel appointed us from our nativity.

17. Spiritus Domini incubabat acquis Gen. 1.

and in six dayes hatched the world; this is that irradiation that dispells the mists of Hell, the clouds of horrour, feare, sorrow, and despaire; and preserves the region of the mind in serenity: whosoever feels not the warme gale and gentle ventilation of this Spirit, (though I feele his pulse) I dare not say he lives; for truely without this, to mee there is no heat under the Tropick; nor any light, though I dwelt in the body of the Sunne.

As when the labouring sun hath wrought his track,
Up to the top of lofty Cancers *back,*
The ycie Ocean cracks, the frozen pole
Thawes with the heat of that celestiall coale;
So when thy absent beames begin t' impart
Againe a Solstice on my frozen heart,
My winters ov'r, my drooping spirits sing,
And every part revives into a Spring.
But if thy quickning beames a while decline,
And with their light blesse not this Orbe of mine,
A chilly frost surpriseth every member,
And in the midst of June I feele December.
Keep still in my horizon, for, to mee,
'Tis not the Sun, that makes the day, but thee.
O how this earthly temper doth debase
The noble Soule, in this her humble place!
Whose wingy nature ever doth aspire
To reach that place whence first it took its fire.
These flames I feele, which in my heart doe dwell,
Are not thy beames, but take their fire from hell:
O quench them all, and let thy light divine
Be as the Sunne to this poore Orbe of mine.
And to thy sacred Spirit convert those fires,
Whose earthly fumes choake my devout aspires.

33. Therefore, for Spirits I am so far from denying their existence, that I could easily beleeve, that not onely whole Countries, but particular

persons have their Tutelary, and Guardian Angels: It is not a new opinion of the Church of *Rome*, but an old one of *Pythagoras* and *Plato*; there is no heresie in it, and if not manifestly defin'd in Scripture, yet is it an opinion of a good and wholesome use in the course and actions of a mans life, and would serve as an *Hypothesis* to salve many doubts, whereof common Philosophy affordeth no solution: Now, if you demand my opinion and Metaphysicks of their natures, I confesse them very shallow, most in a negative way, like that of God; or in a comparative, between our selves and fellow creatures; for there is in this Universe a Staire, or manifest Scale of creatures, rising not disorderly, or in confusion, but with a comely method and proportion: betweene creatures of mere existence and things of life, there is a large disproportion of nature; betweene plants and animals or creatures of sense, a wider difference; between them and man, a farre greater: and if the proportion held on, betweene man and Angels there should bee yet a greater. We doe not comprehend their natures, who retaine the definition[18] of *Porphyry*, and distinguish them from our selves by immortality; for before his fall, man also was immortall; yet must wee needs affirme that he had a different essence from the Angels: having therefore no certaine knowledge of their natures, 'tis no bad method of the Schooles, whatsoever perfection we finde obscurely in our selves, in a more compleate and absolute way to ascribe unto them. I beleeve they have an extemporary knowledge, and upon the first motion of their reason doe what we cannot without study or deliberation; that they know things by their formes, and define by specificall difference,* what we describe by accidents and properties; and therefore probabilities to us may bee demonstrations unto them; that they have knowledge not onely of the specificall, but of the numericall formes of individualls, and understand by what reserved difference* each single *Hypostasis** (besides the relation to its species) becomes its numericall selfe. That as the Soule hath a power to move the body it informes, so there's a Faculty to move any, though informe none; ours upon restraint of

18. Essentia rationalis immortalis.

time, place, and distance; but that invisible hand that conveyed *Habakkuk* to the Lions den,* or *Philip* to *Azotus*,* infringeth this rule, and hath a secret conveyance, wherewith mortalitie is not acquainted; if they have that intuitive knowledge, whereby as in reflection they behold the thoughts of one another, I cannot peremptorily deny but they know a great part of ours. They that to refute the Invocation of Saints, have denied that they have any knowledge of our affaires below, have proceeded too farre, and must pardon my opinion, till I can thoroughly answer that piece of Scripture, *At the conversion of one sinner the Angels of heaven rejoyce.* I cannot with that great Father* securely interpret the worke of the first day, *Fiat lux*,* to the creation of Angels, though (I confesse) there is not any creature that hath so neare a glympse of their nature, as light in the Sunne and Elements; we stile it a bare accident, but where it subsists alone, 'tis a spirituall Substance, and may bee an Angel: in briefe, conceive light invisible, and that is a Spirit.

34. These are certainly the Magisteriall & master pieces of the Creator, the Flower (or as we may say) the best part of nothing, actually existing, what we are but in hopes, and probabilitie; we are onely that amphibious piece betweene a corporall and spirituall essence, that middle frame that linkes those two together, and makes good the method of God and nature, that jumps not from extreames, but unites the incompatible distances by some middle and participating natures; that wee are the breath and similitude of God, it is indisputable, and upon record of holy Scripture; but to call our selves a Microcosme, or little world, I thought it onely a pleasant trope of Rhetorick, till my nearer judgement and second thoughts told me there was a reall truth therein: for first wee are a rude masse, and in the ranke of creatures, which only are, and have a dull kinde of being, not yet priviledged with life, or preferred* to sense or reason; next we live the life of plants, the life of animals, the life of men, and at last the life of spirits, running on in one mysterious nature those five kinds of existences, which comprehend the creatures not of the world, onely, but of the Universe; thus is man that great and true *Amphibium*, whose nature is disposed to live not onely like other

creatures in divers elements, but in divided and distinguished worlds; for though there bee but one world to sense, there are two to reason; the one visible; the other invisible, whereof *Moses* seemes to have left no description; and of the other* so obscurely that some parts[19] thereof are yet in controversie; and truely for those first chapters of *Genesis*, I must confesse a great deale of obscurity, though Divines have to the power of humane reason endeavoured to make all goe in a literall meaning; yet those allegoricall interpretations are also probable, and perhaps the mysticall method of *Moses* bred up in the Hieroglyphicall Schooles of the Egyptians.

35. Now for that immateriall world, me thinkes we need not wander so farre as the first moveable;* for even in this materiall fabricke the spirits walke as freely exempt from the affections of time, place, and motion, as beyond the extreamest circumference; doe but extract from the corpulency of bodies, or resolve things beyond their first matter, and you discover the habitation of Angels, which if I call the ubiquitary, and omnipresent essence of God, I hope I shall not offend Divinity; for before the Creation of the world God was really all things. For the Angels hee created no new world, or determinate mansion, and therefore they are every where his essence is and doe live, at a distance, even in himselfe: that God made all things for man, is in some sense true; yet not so farre as to subordinate the creation of those purer creatures unto ours, though as ministring spirits they doe, and are willing to fulfill the will of God in these lower and sublunary affaires of man; God made all things for himself, and it is impossible hee should make them for any other end than his owne glory; it is all he can receive, and all that is without himselfe; for honour being an externall adjunct, and in the honourer rather than in the person honoured, it was necessary to make a creature, from whom hee might receive this homage, and that is in the other world Angels, in this, man; which when we neglect, we forget the very end of our creation, and may justly provoke God, not onely to repent that hee hath made the world, but that hee hath

19. The Element of fire.

sworne hee would not destroy it. That there is but one world, is a conclusion of faith. *Aristotle* with all his Philosophy hath not beene able to prove it, and as weakely that the world was eternall; that dispute much troubled the pennes of the antient Philosophers who saw no further than the first matter; but *Moses* hath decided that question, and all is salved with the new terme of a creation, that is, a production of something out of nothing; and what is that? Whatsoever is opposite to something or more exactly, that which is truely contrary unto God: for he onely is, all others have an existence with dependency and are something but by a distinction; and herein is Divinity conformant unto Philosophy, and generation not onely founded on contrarieties, but also creation; God being all things is contrary unto nothing out of which were made all things, and so nothing became something, and *Omneity* informed *Nullity** into an essence.

36. The whole Creation is a mystery, and particularly that of man; at the blast of his mouth were the rest of the creatures made, and at his bare word they started out of nothing: but in the frame of man (as the text describes it) he played the sensible operator, and seemed not so much to create, as make him; when hee had separated the materials of other creatures, there consequently resulted a forme and soule; but having raised the wals of man, he was driven to a second and harder creation of a substance like himselfe, an incorruptible and immortall soule. For these two affections* we have the Philosophy, and opinion of the Heathens, the flat affirmative of *Plato*, and not a negative from *Aristotle:* there is another scruple cast in by Divinity (concerning its production) much disputed in the *Germane* auditories,* and with that indifferency and equality of arguments, as leave the controversie undetermined. I am not of *Paracelsus* minde that boldly delivers a receipt to make a man without conjunction;* yet cannot but wonder at the multitude of heads that doe deny traduction,* having no other argument to confirme their beliefe, then that Rhetoricall sentence, and *Antimetathesis*[20] of *Augustine, Creando*

20. A figure in Rethoricke where one word is inverted upon another.*

infunditur, infundendo creatur: either opinion will consist well enough with religion; yet I should rather incline to this, did not one objection haunt mee; not wrung from speculations and subtilties, but from common sense, and observation; not pickt from the leaves of any author, but bred among the weeds and tares of mine owne braine. And this is a conclusion from the equivocall and monstrous productions in the copulation of man with beast; for if the soule of man bee not transmitted and transfused in the seed of the parents, why are not those productions meerely beasts, but have also an impression and tincture of reason in as high a measure as it can evidence it selfe in those improper organs? Nor truely can I peremptorily deny, that the soule in this her sublunary estate, is wholly and in all acceptions inorganicall; but that for the performance of her ordinary actions, is required not onely a symmetry and proper disposition of Organs, but a Crasis* and temper correspondent to its operations; yet is not this masse of flesh and visible structure the instrument and proper corps of the soule, but rather of sense, and that the hand of reason. In our study of Anatomy there is a masse of mysterious Philosophy, and such as reduced the very Heathens to Divinitie; yet amongst all those rare discoveries, and curious pieces I finde in the fabricke of man, I doe not so much content my selfe as in that I finde not, that is, no Organ or instrument for the rationall soule; for in the braine, which we tearme the seate of reason, there is not any thing of moment more than I can discover in the cranie of a beast: and this is a sensible and no inconsiderable argument of the inorganity of the soule, at least in that sense we usually so receive it. Thus are we men, and we know not how; there is something in us, that can be without us, and will be after us, though it is strange that it hath no history, what it was before us, nor can tell how it entred in us.

37. Now for the wals of flesh, wherein the soule doth seeme to be immured before the Resurrection, it is nothing but an elementall composition, and a fabricke that must fall to ashes; *All flesh is grasse*, is not onely metaphorically, but literally true, for all those creatures which we behold, are but the hearbs of the field, digested into flesh in them, or more remotely carnified* in our selves. Nay further, we

are what we all abhorre, *Antropophagi* and Cannibals, devourers not onely of men, but of our selves; and that not in an allegory, but a positive truth; for all this masse of flesh which wee behold, came in at our mouths: this frame wee looke upon, hath beene upon our trenchers; In briefe, we have devoured our selves and yet do live and remaine our selves. I cannot beleeve the wisedome of *Pythagoras* did ever positively, and in a literall sense, affirme his *Metempsychosis*, or impossible transmigration of the soules of men into beasts: of all Metamorphoses and transformations, I beleeve onely one, that is of *Lots* wife,* for that of *Nabuchodonosor** proceeded not so farre; In all others I conceive there is no further verity then is contained in their implicite sense and morality: I beleeve that the whole frame of a beast doth perish, and is left in the same state after death, as before it was materialled unto life; that the soules of men know neither contrary nor corruption, that they subsist beyond the body, and outlive death by the priviledge of their proper natures, and without a miracle; that the soules of the faithfull as they leave earth, take possession of Heaven: that those apparitions, and ghosts of departed persons, are not the wandring soules of men, but the unquiet walkes of Devils, prompting and suggesting us unto mischiefe, bloud, and villany; instilling, & stealing into our hearts, that the blessed spirits are not at rest in their graves, but wander solicitous of the affaires of the world. That those phantasmes appeare often, and doe frequent Cemiteries, charnall houses, and Churches, it is because those are the dormitories of the dead, where the Devill like an insolent Champion beholds with pride the spoyles and Trophies of his victory in *Adam*.

38. This is that dismall conquest we all deplore, that makes us so often cry (O) *Adam, quid fecisti?** I thanke God I have not those strait ligaments, or narrow obligations to the world, as to dote on life, or be convulst and tremble at the name of death: Not that I am insensible of the dread and horrour thereof, or by raking into the bowells of the deceased, or the continual sight of Anatomies, Skeletons, or Cadaverous reliques, like Vespilloes* or Grave-makers, I am become stupid, or have forgot the apprehension of mortality; but that marshalling all the horrours, and contemplating the extremities

thereof, I finde not any thing therein able to daunt the courage of a man, much lesse a well resolved Christian. And therefore am not angry at the errour of our first parents, or unwilling to beare a part in this common fate, and like the best of them to dye; that is, to cease to breathe, to take a farewell of the elements, to be a kinde of nothing for a moment, to be within one instant a spirit. When I take a full view and circle of my selfe, without this reasonable moderator, and equal piece of justice, Death, I doe conceive my selfe the miserablest person extant; were there not another life that I hope for, all the vanities of this world should not intreat a moments breath from me; could the Devill worke my beliefe to imagine I could never dye, I would not out-live that very thought; I have so abject a conceit of this common way of existence, this retaining to the Sunne and Elements, I cannot thinke this is to be a man, or to live according to the dignitie of humanity; in expectation of a better I can with patience embrace this life; yet in my best meditations doe often desire death; It is a symptom of melancholy to be afraid of death, yet sometimes to desire it; this latter I have often discovered in my selfe, and thinke no man ever desired life as I have sometimes death. I honour any man that contemnes it, nor can I highly love any that is afraid of it; this makes me naturally love a Souldier, and honour those tattered and contemptible Regiments that will die at the command of a Sergeant. For a Pagan there may bee some motives to bee in love with life; but for a Christian that is amazed at death, I see not how hee can escape this Dilemma: that he is too sensible of this life, or hopelesse of the life to come.

39. Some Divines count *Adam* 30 yeares old at his creation, because they suppose him created in the perfect age and stature of man; and surely wee are all out of the computation of our age, and every man is some moneths elder than hee bethinkes him; for we live, move, have a being, and are subject to the actions of the elements, and the malice of diseases in that other world, the truest Microcosme, the wombe of our mother; for besides that generall and common existence wee are conceived to hold in our Chaos, and whilst wee sleepe within the bosome of our causes, wee enjoy a being

and life in three distinct worlds, wherein we receive most manifest graduations: In that obscure world and wombe of our mother, our time is short, computed by the Moone, yet longer than the dayes of many creatures that behold the Sunne; our selves being yet not without life, sense, and reason; though for the manifestation of its actions it awaits the opportunity of objects; and seemes to live there but in its roote and soule of vegetation: entring afterwards upon the scene of the world, wee arise up and become another creature, performing the reasonable actions of man, and obscurely manifesting that part of Divinity in us, but not in complement and perfection, till we have once more cast our secondine,* that is, this slough of flesh, and are delivered into the last world, that ineffable place of Paul, that proper *ubi** of spirits. The smattering I have in the knowledge of the Philosophers stone, (which is something more then the perfect exaltation* of gold) hath taught me a great deale of Divinity, and instructed my beliefe, how that immortall spirit and incorruptible substance of my soule may lye obscure, and sleepe a while within this house of flesh. Those strange and mysticall transmigrations that I have observed in Silkewormes, turn'd my Philosophy into Divinity. There is in those workes of nature, which seeme to puzle reason, something Divine, and that hath more in it then the eye of a common spectator doth discover. I have therefore forsaken those strict definitions of death, by privation of life, extinction of naturall heate, separation etc. of soule and body, and have framed one in an hermeticall way unto mine owne fancie: *est mutatio qua perficitur nobile illud extractum Microcosmi*;* for to mee that consider things in a naturall and experimentall way, man seemes to be but a digestion, or a preparative way unto that last and glorious Elixar which lies imprisoned in the chaines of flesh, etc..

40. I am naturally bashfull, nor hath conversation, age, or travell, beene able to effront, or enharden me; yet I have one part of modesty, which I have seldome discovered in another, that is (to speake truly) I am not so much afraid of death, as ashamed thereof; tis the very disgrace and ignominy of our natures, that in a moment can so disfigure us that our nearest friends, wives and Children stand

afraid and start at us. The Birds and Beasts of the field that before in a naturall feare obeyed us, forgetting all allegiance begin to prey upon us. This very conceite hath in a tempest disposed and left me willing to be swallowed in the abysse of waters; wherein I had perished unseene, unpityed, without wondring eyes, teares of pity, Lectures of mortality, and none had said, *quantum mutatus ab illo!** Not that I am ashamed of the Anatomy of my parts, or can accuse nature for playing the bungler in any part of me, or my owne vitious life for contracting any shamefull disease upon me, whereby I might not call my selfe as wholesome a morsell for the wormes as any.

41. Some upon the courage of a fruitfull issue, wherein, as in the truest Chronicle, they seem to outlive themselves, can with greater patience away with death. This conceit and counterfeit subsisting in our progenies seemes to mee a meere fallacy, unworthy the desires of a man, that can but conceive a thought of the next world; who, in a nobler ambition, should desire to live in his substance in Heaven rather than in his name and shadow on earth. And therefore at my death I meane to take a totall adieu of the world, not caring for a Monument, History, or Epitaph, not so much as the bare memory of my name to be found any where but in the universall Register of God: I am not yet so Cynicall, as to approve the Testament[21] of *Diogenes*, nor doe I altogether allow that *Rodomontado** of *Lucan*;

Cælo tegitur, qui non habet urnam.
He that unburied lies wants not his Herse,
For unto him a tombe's the Universe.

But commend in my calmer judgement, those ingenuous intentions that desire to sleepe by the urnes of their Fathers, and strive to goe the nearest way unto corruption. I do not envie the temper of Crowes and Dawes, nor the numerous and weary dayes of our Fathers before the Flood. If there bee any truth in Astrology, I may outlive a

21. Who willed his friend not to bury him but to hang him up with a staffe in his hand to fright away the crowes.

Jubilee;[22] as yet I have not seene one revolution of *Saturne*,[23] nor hath my pulse beate thirty yeares, and yet, excepting one, have seene the Ashes, and left under ground all the Kings of *Europe*, have beene contemporary to three Emperours, foure Grand Signiours, and as many Popes; mee thinkes I have outlived my selfe, and begin to bee weary of the Sunne; I have shaken hands with delight in my warme blood and Canicular dayes;* I perceive I doe Anticipate the vices of age, the world to mee is but a dreame, or mockshow, and wee all therein but Pantalones[24] and Antickes to my severer contemplations.

42. It is not, I confesse, an unlawfull Prayer to desire to surpasse the dayes of our Saviour, or wish to out-live that age wherein he thought fittest to dye; yet if (as Divinity affirmes) there shall be no gray hayres in Heaven, but all shall rise in the perfect state of men, we doe but out-live those perfections in this world, to be recalled unto them by a greater miracle in the next, and run on here but to be retrograde hereafter. Were there any hopes to out-live vice, or a point to be super-annuated from sin, it were worthy of our knees to implore the dayes of *Methuselah*. But age doth not rectifie, but incurvate* our natures, turning bad dispositions into worser habits, and (like diseases) brings on incurable vices; for every day as we grow weaker in age, we grow stronger in sinne, and the number of our dayes doth but make our sinnes innumerable. The same vice committed at sixteene, is not the same, though it agree in all other circumstances, at forty; but swels and doubles from the circumstance of our ages, wherein besides the constant and inexcusable habit of transgressing, the maturity of our Judgement cuts off pretence unto excuse or pardon: every sin, the oftner it is committed, the more it acquireth in the quality of evill; as it succeeds in time, so it proceeds in degrees of badnesse; for as they proceed they ever multiply, and like figures in Arithmeticke, the last stands for more than all that

---

22. The Jewish computation for 50 yeares.

23. The Planet of Saturne makes his revolution once in 30 yeares.

24. A French word for Antickes.

went before it: And though I thinke no man can live well once but hee that could live twice, yet, for my owne part, I would not live over my houres past, or beginne againe the thred of my dayes: not upon *Cicero*'s ground, because I have lived them well, but for feare I should live them worse; I find my growing Judgement dayly instructs me how to be better, but my untamed affections and confirmed vitiosity* make mee dayly doe worse; I finde in my confirmed age the same sinnes I discovered in my youth; I committed many then because I was a child, and because I commit them still I am yet an Infant. Therefore I perceive a man may bee twice a child before the dayes of dotage, and stand in need of *Aesons* bath* before threescore.

43. And truely there goes a great deale of providence to produce a mans life unto threescore; there is more required than an able temper for those yeeres; though the radicall humour* containe in it sufficient oyle for seventie, yet I perceive in some it gives no light past thirtie; men assigne not all the causes of long life that write whole books thereof. They that found themselves on the radicall balsome or vitall sulphur* of the parts, determine not why *Abel* liv'd not so long as *Adam*. There is therefore a secret glome* or bottome of our dayes; 'twas his wisedome to determine them, but his perpetuall and waking providence that fulfils and accomplisheth them, wherein the spirits, our selves, and all the creatures of God in a secret and disputed way doe execute his will. Let them not therefore complaine of immaturitie that die about thirty; they fall but like the whole world, whose solid and well composed substance must not expect the duration and period of its constitution; when all things are compleated in it, its age is accomplished, and the last and generall fever may as naturally destroy it before six thousand, as me before forty: there is therfore some other hand that twines the thread of life than that of nature; wee are not onely ignorant in Antipathies and occult qualities, our ends are as obscure as our beginnings, the line of our dayes is drawne by night, and the various effects therein by a pencill that is invisible; wherein though wee confesse our ignorance, I am sure we doe not erre, if wee say, it is the hand of God.

44. I am much taken with two verses of *Lucan*, since I have beene able not onely, as we doe at Schoole, to construe, but understand them:

> *Victurosque Dei celant ut vivere durent,*
> *Felix esse mori.*
> *We're all deluded, vainely searching wayes,*
> *To make us happy by the length of dayes;*
> *For cunningly to make's protract this breath,*
> *The Gods conceale the happiness of Death.*

There be many excellent straines in that Poet, wherewith his Stoicall Genius hath liberally supplyed him; and truely there are singular pieces in the Philosophy of *Zeno*, and doctrine of the Stoickes, which I perceive, delivered in a Pulpit, passe for currant Divinity: yet herein are they in extreames, that can allow a man to be his owne *Assassine*, and so highly extoll the end and suicide of *Cato*; this is indeed not to feare death, but yet to bee afraid of life. It is a brave act of valour to contemne death, but where life is more terrible than death, it is then the truest valour to dare to live, and herein Religion hath taught us a noble example: For all the valiant acts of *Curtius*, *Scevola* or *Codrus*, do not parallel or match that one of *Job*; and surely there is no torture to the racke of a disease, nor any Poynyards in death it selfe like those in the way or prologue unto it. *Emori nolo, sed me esse mortuum nihil curo,** I would not die, but care not to be dead. Were I of *Cæsars* Religion I should be of his desires, and wish rather to goe off at one blow, then to be sawed in peeces by the grating torture of a disease. Men that looke no further than their outsides thinke health an appertinance unto life, and quarrell with their constitutions for being sick; but I that have examined the parts of man, and know upon what tender filaments that Fabrick hangs, doe wonder that we are not alwayes so; and considering the thousand dores that lead to death doe thanke my God that we can die but once. 'Tis not onely the mischiefe of diseases, and the villanie of poysons that make an end of us; we vainly accuse the fury of Gunnes, and the new inventions of death; 'tis in the power of every hand to destroy us, and wee are beholding unto every one wee meete hee doth not kill us. There is therefore but one comfort left, that though it be in the power of the weakest arme to take away life, it is not in the strongest to deprive us of death: God would not exempt himselfe from that; the misery of immortality in the flesh he undertooke not, that was in it immortall. Certainly there is no happinesse within this circle of flesh, nor is it in the Opticks of these eyes to behold felicity; the first day of our Jubilee is death; the devill hath therefore fail'd of his desires; wee are happier with death than we should have beene without it: there is no misery but in himselfe where there is no end of misery; and so indeed in his owne sense, the Stoick is in the right: Hee forgets that hee can die who complaines of misery, wee are in the power of no calamitie, while death is in our owne.

45. Now besides this literall and positive kinde of death, there are others whereof Divines make mention, and those I thinke, not meerely Metaphoricall, as Mortification, dying unto sin and the world; therefore, I say, every man hath a double Horoscope, one of his humanity, his birth; another of his Christianity, his baptisme, and from this doe I compute or calculate my Nativitie; not reckoning those *Horæ combustæ*,[25] and odde dayes, or esteeming my selfe any thing, before I was my Saviours, and inrolled in the Register of Christ: Whosoever enjoyes not this life, I count him but an apparition, though he weare about him the sensible affections of flesh. In these morall acceptions, the way to be immortall is to die daily; nor can I thinke I have the true Theory of death, when I contemplate a skull, or behold a Skeleton with those vulgar imaginations it casts upon us; I have therefore enlarged that common *Memento mori*, into a more Christian memorandum, *Memento quatuor novissima*,* those foure inevitable points of us all, Death, Judgement, Heaven and Hell. Neither did the contemplations of the Heathens rest in

25. That time when the moone is in conjunction, and obscured by the Sun, the Astrologers call *Horæ Combustæ*.

their graves, without a further thought of *Rhadamanth* or some judiciall proceeding after death, though in another way, and upon suggestion of their naturall reason. I cannot but marvaile from what *Sybill* or Oracle they stole the prophesy of the worlds destruction by fire, or whence *Lucan** learned to say,

Communis mundo superest rogus, ossibus astra Misturus.
There yet remaines to th' world one common fire,
Wherein our bones with stars shall make one pyre.

I do beleeve the world drawes near its end, yet is neither old nor decayed, nor will ever perish upon the ruines of its owne principles. As the Creation was a worke above nature, so is its adversary, annihilation; without which the world hath not its end, but its mutation. Now what fire should bee able to consume it thus farre, without the breath of God, which is the truest consuming flame, my Philosophy cannot informe me. Some beleeve there went not a minute to the worlds creation, nor shal there go to its destruction; those six dayes so punctually described, make not to them one moment, but rather seem to manifest the method and Idea of that great worke in the intellect of God, than the manner how hee proceeded in its operation. I cannot dream that there should be at the last day any such Judiciall proceeding, or calling to the Barre, as indeed the Scripture seemes to imply, and the literall commentators doe conceive: for unspeakable mysteries in the Scriptures are often delivered in a vulgar and illustrative way, and being written unto man, are delivered, not as they truely are, but as they may bee understood; wherein, notwithstanding the different interpretations according to the different capacities, they may stand firme with our devotion, nor bee any way prejudiciall to each single edification.

46. Now to determine the day and yeare of this inevitable time, is not onely convincible* and statute madnesse, but also manifest impiety; How shall we interpret *Elias* 6000 yeares,* or imagine the secret communicated to a Rabbi, which a God hath denyed unto his Angels? It had beene an excellent quære,* to have posed the devill of

Delphos[26] and must needs have forced him to some strange amphibology;* it hath not onely mocked the predictions of sundry Astrologers in ages past, but the prophecies of many melancholy heads in these present, who neither reasonably understanding things past, nor present, pretend a knowledge of things to come; heads ordained onely to manifest the incredible effects of melancholy, and to fulfill old prophesies, rather than be the authors of new; In those dayes there shall be warres and rumours of warres,[27] to me seemes no prophesie, but a constant truth, in all times verified since it was first pronounced: There shall bee signes in the Moone and Starres; how comes he then like a theefe in the night, when he gives an item of his comming? That common signe drawne from the revelation of Antichrist is as obscure as any; in our common compute he hath beene come these many yeares, but for my owne part to speake freely, omitting those ridiculous Anagrams,[28] I am half of Paracelsus opinion, and thinke Antichrist the Philosophers stone in Divinity, for the discovery and invention whereof, though there be prescribed rules, and probable inductions, yet hath hardly any man attained the perfect discovery thereof. That generall opinion that the world drawes neere its end, hath possessed all ages past as neerely as ours; I am afraid that the Soules that now depart, cannot escape that lingring expostulation of the Saints under the Altar, *Quousque Domine? How long, O Lord?* and groane in the expectation of the great Jubilee.

47. This is the day that must make good that great attribute of God, his Justice; that must reconcile those unanswerable doubts which torment the wisest understandings, and reduce those seeming inequalities, and respective* distributions in this world, to an equality and recompensive Justice in the next. This is that one day, that shall include and comprehend all that went before it, wherein,

---

26. The Oracle of Apollo.

27. In those dayes there shall come lyers and false prophets.

28. Whereby men labour to prove the Pope Antichrist, from their names making up the number of the Beast.

as in the last scene, all the Actors must enter to compleate and make up the Catastrophe of this great peece. This is the day whose memory hath onely power to make us honest in the darke, and to bee vertuous without a witnesse. *Ipsa sui pretium virtus sibi*, that vertue is her owne reward, is but a cold principle, and not able to maintaine our variable resolutions in a constant and setled way of goodnesse. I have practised that honest artifice of *Seneca*, and in my retired and solitary imaginations, to detaine me from the foulenesse of vice, have fancyed to my selfe the presence of my deare and worthiest friends, before whom I should lose my head, rather than be vitious; yet herein I found that there was nought but morall honesty, and this was not to be vertuous for his sake who must reward us at the last. I have tryed if I could reach that great resolution of his,* to be honest without a thought of Heaven or Hell; and indeed I found upon a naturall inclination, and inbred loyalty unto vertue, that I could serve her without a livery, yet not in that resolved and venerable way, but that the frailty of my nature, upon an easie temptation, might be induced to forget her. The life therefore and spirit of all our actions, is the resurrection, and stable apprehension, that our ashes shall enjoy the fruits of our pious endeavours; without this, all Religion is a Fallacy, and those impieties of *Lucian*, *Euripides*, and *Julian* are no blasphemies, but subtile verities, and Atheists have beene the onely Philosophers.

48. How shall the dead arise, is no question of my faith; to beleeve onely possibilities, is not faith, but meere Philosophy; many things are true in Divinity, which are neither inducible by reason, nor confirmable by sense; and many things in Philosophy confirmable by sense, yet not inducible by reason. Thus it is impossible by any solid or demonstrative reasons to perswade a man to beleeve the conversion of the Needle to the North; though this be possible, and true, and easily credible, upon a single experiment unto the sense. I beleeve that our estranged and divided ashes shall unite againe; that our separated dust after so many pilgrimages and transformations into the parts of mineralls, Plants, Animals, Elements, shall at the voyce of God returne into their primitive shapes, and joyne againe

to make up their primary and predestinated formes. As at the Creation of the world, there was a separation of that confused masse into its species, so at the destruction thereof there shall bee a separation into its distinct individuals. As at the Creation of the world, all the distinct species that wee behold, lay involved in one masse, till the fruitfull voyce of God separated this united multitude into its several species: so at the last day, when these corrupted reliques shall be scattered in the wildernesse of formes, and seeme to have forgot their proper habits, God by a powerfull voyce shall command them backe into their proper shapes, and call them out by their single individuals: Then shall appeare the fertilitie of *Adam*, and the magicke of that sperme that hath dilated into so many millions. What is made to be immortall, Nature cannot, nor will the voyce of God, destroy. Those bodies that we behold to perish, were in their created nature immortall, and liable unto death but accidentally and upon forfeit; and therefore they owe not that naturall homage unto death as other bodies doe, but may be restored to immortality with a lesser miracle and by a bare and easie revocation of course return immortall. I have often beheld as a miracle that artificiall resurrection and revivification of *Mercury*, how being mortified* into a thousand shapes, it assumes againe its owne, and returns to its numericall selfe. Let us speake naturally, and like Philosophers: the formes of alterable bodies in these sensible corruptions perish not; nor, as wee imagine, wholly quit their mansions, but retire and contract themselves into their secret and inaccessible parts, where they may best protect themselves from the action of their Antagonist. A plant or vegetable consumed to ashes, to a contemplative and schoole Philosopher seemes utterly destroyed, and the forme to have taken his leave for ever: But to a sensible Artist* the formes are not perished, but withdrawne into their incombustible part, where they lie secure from the action of that devouring element. This is made good by experience, which can from the ashes of a plant revive the plant, and from its cinders recall it into its stalk and leaves againe. What the Art of man can doe in these inferiour pieces, what blasphemy is it to affirme the finger of God cannot doe in these more perfect and sensible structures? This is that mysticall Philosophy, from whence no true Scholler becomes an Atheist, but from the visible effects of nature, growes up a reall Divine, and beholds not in a dreame, as *Ezekiel*, but in an ocular and visible object the types* of his resurrection.

49. Now, the necessary Mansions of our restored selves are those two contrary and incompatible places wee call Heaven and Hell; for to define them, or strictly to determine what and where they are, surpasseth my Divinity. That elegant Apostle* which seemed to have a glimpse of Heaven, hath left but a negative description thereof; Which neither eye hath seen, nor eare hath heard, nor can enter into the heart of man: he was translated out of himself to behold it, but being returned into himself could not expresse it. Saint *Johns* description by Emeralds Chrysolites, and pretious stones, is too weake to expresse the materiall Heaven we behold. Briefely therefore, where the soule hath the full measure, and complement of happinesse, where the boundlesse appetites of that spirit remaine compleatly satisfied, that it cannot desire either addition or alteration, that I thinke is truely Heaven: and this can onely be in the enjoyment of that essence, whose infinite goodnesse is able to terminate the desires of it selfe, and the insatiable wishes of ours; wherever God will thus manifest himselfe, there is Heaven, though within the circle of this sensible world. Thus the soule of man may bee in Heaven any where, even within the limits of his owne proper body, and when it ceaseth to live in the body, it may remaine in its owne soule, that is its Creator. And thus wee may say that Saint *Paul*, whether in the body or out of the body, was yet in Heaven. To place it in the Empyreall,* or beyond the tenth Spheare, is to forget the worlds destruction; for when this sensible world shall bee destroyed, all shall then be here as it is now there, an Empyreall Heaven, a *quasi* vacuitie or place exempt from the naturall affection of bodies, where to aske where Heaven is, is to demand where the presence of God is, or where wee have the glory of that happy vision. *Moses* that was bred up in all the learning of the *Egyptians*, committed a grosse absurdity in Philosophy, when with these eyes of flesh he desired to see God, and petitioned his Maker, that is truth it selfe, to a contradiction.

Those that imagine Heaven and Hell neighbours, and conceive a vicinity between those two extreames, upon consequence of the Parable,* where *Dives* discoursed with *Lazarus* in *Abrahams* bosome, do too grossely conceive of those glorified creatures,* whose eyes shall easily out-see the Sunne, and behold without a Perspective,* the extremest distances: for if there shall be in our glorified eyes, the faculty of sight & reception of obiects, I could thinke the visible species* there to be in as unlimitable a way as now the intellectuall. I grant that two bodies placed beyond the tenth Spheare, or in a vacuity, according to *Aristotles* Philosophy, could not behold each other, because there wants a body or Medium to hand and transport the visible rayes of the object unto the sense; but when there shall be a generall defect of either Medium to convey, or light to prepare & dispose that Medium, and yet a perfect vision, wee must suspend the rules of our Philosophy, and make all good by a more absolute piece of Opticks.

50. I cannot tell how to say that fire is the essence of hell; I know not what to make of Purgatory, or conceive a flame that can either prey upon, or purifie the substance of a soule; those flames of sulphure mentioned in the Scriptures, I take to be understood not of this present Hell, but of that to come, where fire shall make up the complement of our tortures, & have a body or subject wherein to manifest its tyranny: Some who have had the honour to be textuaries* in Divinity, are of opinion it shall be the same specificall fire with ours. This is hard to conceive, yet can I make good how even that may prey upon our bodies, and yet not consume us: for in this materiall world, there are bodies that persist invincible in the powerfullest flames, and though by the action of fire they fall into ignition and liquation, yet will they never suffer a destruction: I would gladly know how *Moses* with an actuall fire calcin'd,[29] or burnt the golden Calfe into powder: for that mysticall mettle of gold, whose solary and celestiall nature I admire, exposed unto the violence of fire, grows onely hot and liquifies, but consumeth not either in its sub-

29. Calcination, a Chymicall terme for the reduction of a minerall into Powder.

stance, weight or vertue: so when the consumable and volatile pieces of our bodies shall be refined into a more impregnable and fixed temper like gold, though they suffer from the action of flames, they shall never perish, but lie immortall in the armes of fire. And surely if this frame must suffer onely by the action of this element, there will many bodies escape, and not onely Heaven, but earth will not bee at an end, but rather a beginning; For at present it is not earth, but a composition of fire, water, earth, and aire; but at that time, spoyled of these ingredients, it shall appeare in a substance more like it selfe, its ashes. Philosophers that opinioned the worlds destruction by fire, did never dreame of annihilation, which is beyond the power of sublunary causes; for the last and powerfullest action of that element is but vitrification or a reduction of a body into Glasse; & therefore some of our Chymicks facetiously affirm, yea, and urge Scripture for it, that at the last fire all shall be crystallized & reverberated* into glasse, which is the utmost action of that element. Nor need we fear this term of annihilation or wonder that God will destroy the workes of his Creation: for man subsisting, who is, and will then truely appeare a Microcosme, the world cannot bee said to be destroyed. For the eyes of God, and perhaps also of our glorified senses, shall as really behold and contemplate the world in its Epitome or contracted essence, as now they doe at large and in its dilated substance. In the seed of a Plant to the eyes of God, and to the understanding of man, there exist, though in an invisible way, the perfect leaves, flowers, and fruits thereof: (for things that are in *posse** to the sense, are actually existent to the understanding). Thus God beholds all things, who contemplates as fully his workes in their Epitome, as in their full volume, and beheld as amply the whole world in that little compendium of the sixth day, as in the scattered and dilated pieces of those five before.

51. Men commonly set forth the tortures of Hell by fire, and the extremitie of corporall afflictions, and describe Hell in the same manner as *Mahomet* doth Heaven. This indeed makes a noyse, and drums in popular eares: but if this be the terrible piece thereof, it is not worthy to stand in diameter with Heaven, whose happinesse

consists in that part which is best able to comprehend it, that immortall essence, that translated divinity and colony of God, the soule. Surely though wee place Hell under earth, the Devils walke and purlue is about it; men speake too popularly who place it in those flaming mountaines, which to grosser apprehensions represent Hell. The heart of man is the place the devill dwels in; I feele somtimes a hell within my selfe, *Lucifer* keeps his court in my brest, Legion* is revived in me. There are as many hels as *Anaxagoras* conceited* worlds; there was more than one hell in *Magdalen*, when there were seven devils; for every devill is an hell unto himselfe: hee holds enough of torture in his owne *ubi*, and needs not the misery of circumference to afflict him, and thus a distracted conscience here is a shadow or introduction unto hell hereafter; Who can but pity the mercifull intention of those hands that doe destroy themselves? the devill, were it in his power, would doe the like; which being impossible, his miseries are endlesse, and he suffers most in that attribute, wherein he is impassible,* his immortality.

52. I thanke God, and with joy I mention it, I was never afraid of Hell, nor ever grew pale at the description of that place; I have so fixed my contemplations on Heaven, that I have almost forgot the Idea of Hell, and am afraid rather to lose the joyes of the one than endure the misery of the other; to be deprived of them is a perfect hell, & needs me thinkes no addition to compleate our afflictions; that terrible terme hath never detained me from sin, nor do I owe any good action to the name thereof: I feare God, yet am not afraid of him, his mercies make me ashamed of my sins, before his judgements afraid thereof: these are the forced and secondary method of his wisedome, and which he useth not but as the last remedy, and upon provocation: a course rather to deterre the wicked, than incite the vertuous to his worship. I can hardly thinke there was ever any scared into Heaven; they goe the surest way to Heaven, who would serve God without a Hell; other Mercenaries that crouch unto him in feare of Hell, though they terme themselves the servants, are indeed but the slaves of the Almighty.

53. And to be true, and speake my soule, when I survey the occur-

rences of my life, and call into account the finger of God, I can perceive nothing but an abysse and masse of his mercies, either in generall to mankind, or in particular to my selfe; and whether out of the prejudice of my affection, or an inverting and partiall conceit of his mercies, I know not, but those which others terme crosses, afflictions, judgements, misfortunes, to me, who enquire farther into them than their visible effects, they both appeare, and in event have ever proved the secret and dissembled favours of his affection. It is a singular piece of wisedome to apprehend truly, and without passion the workes of God, and so well to distinguish his justice from his mercy, as not to miscall those noble attributes of the Allmighty; yet it is likewise an honest piece of Logick so to dispute and argue the proceedings of God, as to distinguish even his judgements into mercies. For God is mercifull unto all, because better to the worst, than the best deserve, and to say he punisheth none in this world, though it be a Paradox, is no absurdity. To one that hath committed murther, if the Judge should onely ordaine a boxe of the eare, it were a madnesse to call this a punishment, and to repine at the sentence, rather than admire the clemency of the Judge. Thus our offences being mortall, and deserving not onely death, but damnation, if the goodnesse of God be content to traverse and passe them over with a losse, misfortune, or disease, what frensie were it to terme this a punishment, rather than an extremitie of mercy, and to groane under the rod of his judgements, rather than admire the Scepter of his mercies? Therefore to adore, honour, and admire him, is a debt of gratitude due from the obligation of our natures, states and conditions; and with these thoughts, he that knowes them best, will not deny that I adore him; that I obtaine Heaven, and the blisse thereof, is accidentall, and not the intended worke of my devotion, it being a felicitie I can neither thinke to deserve, nor scarce in modesty expect. For these two ends of us all, either as rewards or punishments, are mercifully ordained, and disproportionally disposed unto our actions, the one being so far beyond our deserts, the other so infinitely below our demerits.

54. There is no salvation to those that beleeve not in Christ, that

is, say some, since his Nativity, and as Divinity affirmeth, before also; which makes me much apprehend the end of those honest Worthies and Philosophers which died before his Incarnation. It is hard to place those soules in Hell whose worthy lives doe teach us vertue on earth; methinks amongst those many subdivisions of hell, there might have bin one Limbo left for these: What a strange vision will it be to see their poeticall fictions converted into verities, & their imaginary & fancied Furies, into reall Devils? how strange to them will sound the History of *Adam*, when they shall suffer for him they never heard of? when they that derive their Genealogy from the Gods, shall know they are the unhappy issue of sinfull man? It is an insolent part of reason to controvert* the works of God, or question the justice of his proceedings; Could humility teach others, as it hath instructed me, to contemplate the infinite and incomprehensible distance betwixt the Creator and the creature, or did we seriously perpend* that one Simile of Saint *Paul,* Shall the vessell say unto the Potter, why hast thou made me thus?* it would prevent these arrogant disputes of reason, nor would wee argue the definitive sentence of God, either to Heaven or Hell. Men that live according to the right rule and law of reason, live but in their owne kinde, as beasts doe in theirs; who justly obey the prescript of their natures, and therefore cannot reasonably demand a reward of their actions, as onely obeying the naturall dictates of their reason. It will therefore, and must at last appeare, that all salvation is through Christ; which verity I feare these great examples of vertue must confirme, and make it good, how the perfectest actions of earth have no title or claime unto Heaven.

55. Nor truely doe I thinke the lives of these or of any other were ever correspondent, or in all points conformable unto their doctrines; it is evident that *Aristotle* transgressed the rule of his owne Ethicks; the Stoicks that condemne passion, and command a man to laugh in *Phalaris* his Bull,* could not endure without a groane a fit of the stone* or collick. The *Scepticks* that affirmed they knew nothing, even in that opinion confuted themselves, and thought they knew more than all the world besides. *Diogenes* I hold to bee the most vaineglorious man of his time, and more ambitious in refusing all

honours, than *Alexander* in rejecting none. Vice and the Devill put a fallacie upon our reasons, and provoking us too hastily to run from it, entangle and profound* us deeper in it. The Duke of *Venice*, that yearly weds himselfe unto his Sea, by casting therein a ring of Gold, I will not argue of prodigality, because it is a solemnity of good use and consequence in the State. But the Philosopher that threw his money into the Sea to avoyd avarice, was a notorious prodigal. There is no road or ready way to vertue, it is not an easie point of art to disentangle our selves from this riddle, or web of sin: To perfect vertue, as to Religion, there is required a Panoplia or compleat armour, that whilst we lye at close ward against one vice we lye not open to the venue* of another: And indeed wiser discretions that have the thred of reason to conduct them, offend without a pardon; whereas under heads* may stumble without dishonour. There goe so many circumstances to piece up one good action, that it is a lesson to be good, and wee are forced to be vertuous by the booke. Againe, the practice of men holds not an equall pace, yea, and often runnes counter to their Theory; we naturally know what is good, but naturally pursue what is evill: the Rhetoricke wherewith I perswade another cannot perswade my selfe: there is a depraved appetite in us, that will with patience heare the learned instructions of Reason; but yet performe no further than agrees to its owne irregular Humour. In briefe, we are all monsters, that is, a composition of man and beast, wherein we must endeavour to be as the Poets fancy that wise man *Chiron*,[30] that is, to have the Region of Man above that of Beast, and sense to sit but at the feete of reason. Lastly, I doe desire with God, that all, but yet affirme with men, that very few shall know salvation, that the bridge is narrow, the passage straite unto life; yet those who doe confine the Church of God, either to particular Nations, Churches, or Families, have made it far narrower than our Saviour ever meant it.

56. The vulgarity of those judgements that wrap the Church of God in *Strabo*'s cloake* and restraine it unto Europe, seeme to mee as bad Geographers as *Alexander*, who thought hee had conquer'd all

30. Chiron, a Centaure.*

the world when hee had not subdued the halfe of any part thereof: For wee cannot deny the Church of God both in Asia and Africa, if we doe not forget the peregrinations of the Apostles, the death of their Martyrs, the sessions of many, and even in our reformed judgement lawfull councells, held in those parts in the minoritie and nonage of ours: nor must a few differences more remarkable in the eyes of man than perhaps in the judgement of God, excommunicate from heaven one another, much lesse those Christians who are in a manner all Martyrs, maintaining their faith in the noble way of persecution, and serving God in the fire, whereas we honour him but in the Sunshine. 'Tis true we all hold there is a number of Elect and many to be saved; yet take our opinions together, and from the confusion thereof there will be no such thing as salvation, nor shall any one be saved; for first the Church of *Rome* condemneth us, wee likewise them, the Sub-reformists and Sectaries* sentence the Doctrine of our Church as damnable, the Atomist,* or Familist* reprobates all these, and all these them againe. Thus whilst the mercies of God doe promise us heaven, our conceits and opinions exclude us from that place. There must be therefore more than one Saint *Peter*; particular Churches and Sects usurpe the gates of heaven, and turne the key against each other; and thus we goe to heaven against each others wills, conceits and opinions, and, with as much uncharity as ignorance, doe erre I feare in points, not onely of our own, but one anothers salvation.

57. I beleeve many are saved who to man seeme reprobated, and many reprobated who in the opinion and sentence of man, stand elected; there will appeare at the last day, strange, and unexpected examples, both of his justice and his mercy, and therefore to define either is folly in man, and insolency even in the devils; those acute and subtill spirits, in all their sagacity, can hardly divine who shall be saved; which if they could prognosticate their labour were at an end, nor need they compasse the earth, seeking whom they may devoure. Those who upon a rigid application of the Law, sentence *Solomon* unto damnation, condemne not onely him, but themselves, and the whole world; for by the letter, and written Law of God, we are

without exception in the state of death; but there is a prerogative of God, and an arbitrary pleasure above the letter of his owne Law, by which alone wee can pretend unto salvation, and through which *Solomon* might be as easily saved as those who condemne him.

58. The number of those who pretend unto salvation, and those infinite swarmes who thinke to passe through the eye of this Needle, have much amazed me. That name and compellation of *little Flocke*, doth not comfort but deject my devotion, especially when I reflect upon mine owne unworthinesse, wherein, according to my humble apprehension, I am below them all. I beleeve there shall never be an Anarchy in Heaven, but as there are Hierarchies amongst the Angels, so shall there be degrees of priority amongst the Saints. Yet is it (I protest) beyond my ambition to aspire unto the first rankes; my desires onely are, and I shall be happy therein, to be but the last man, and bring up the Rere in Heaven.

59. Againe, I am confident and fully perswaded, yet dare not take my oath of my salvation; I am as it were sure, and do beleeve, without all doubt, that there is such a city as *Constantinople*; yet for me to take my oath thereon, were a kinde of perjury, because I hold no infallible warrant from my owne sense, to confirme me in the certainty thereof. And truely, though many pretend an absolute certainty of their salvation, yet when an humble soule shall contemplate her owne unworthinesse, she shall meete with many doubts and suddainely finde how much wee stand in need of the precept of Saint *Paul*,* *Worke out your salvation with feare and trembling.* That which is the cause of my election, I hold to be the cause of my salvation, which was the mercy, and beneplacit* of God, before I was, or the foundation of the world. *Before Abraham was, I am*, is the saying of Christ;* yet is it true in some sense if I say it of my selfe, for I was not onely before my selfe, but *Adam*, that is, in the Idea of God, and the decree of that Synod held from all Eternity. And in this sense, I say, the world was before the Creation, and at an end before it had a beginning; and thus was I dead before I was alive; though my grave be *England*, my dying place was Paradise, and *Eve* miscarried of mee before she conceiv'd of *Cain*.

60. Insolent zeales that doe decry good workes and rely onely upon faith, take not away merits: for depending upon the efficacy of their faith, they enforce the condition of God, and in a more sophisticall way doe seeme to challenge Heaven. It was decreed by God, that onely those that lapt in the water like dogges, should have the honour to destroy the *Midianites*;* yet could none of those justly challenge, or imagine hee deserved that honour thereupon. I doe not deny, but that true faith, and such as God requires, is not onely a marke or token, but also a meanes of our Salvation; but where to finde this, is as obscure to me, as my last end. And if our Saviour could object unto* his owne Disciples, & favourites, a faith, that to the quantity of a graine of Mustard seed, is able to remove mountaines; surely that which wee boast of, is not anything, or at the most, but a remove from nothing. This is the Tenor of my beleefe, wherein, though there be many things singular, and to the humour of my irregular selfe, yet, if they square not with maturer Judgements, I disclaime them, and doe no further father them than the learned and best Judgements shall authorize them.

THE SECOND PART

1. NOW FOR that other Vertue of Charity, without which Faith is a meer notion, and of no existence, I have ever endeavoured to nourish the mercifull disposition, and humane inclination which I borrowed from my Parents, and to regulate it to the written and prescribed Lawes of Charity; and if I hold the true Anatomy of my selfe, I am delineated & naturally framed to such a piece of vertue: for I am of a constitution so generall, that it consorts and sympathizeth with all things; I have no antipathy, or rather Idio-syncrasie, in dyet, humour, ayre, anything; I wonder not at the *French*, for their dishes of frogges, snailes, and toadstooles, nor at the Jewes for Locusts and Grasse-hoppers, but being amongst them, make them my common viands; and I finde they agree with my stomach as well as theirs; I could digest a Sallad gathered in a Church-yard, as well as in a Garden. I cannot start at the presence of a Serpent, Scorpion, Lizard, or Salamander; at the sight of a Toad, or Viper, I feel in me no desire to take up a stone to destroy them. I finde not in my selfe those common antipathies that I can discover in others: Those nationall repugnances doe not touch me, nor doe I behold with prejudice the *French*, *Italian*, *Spaniard*, or *Dutch*; but where I finde their actions in ballance with my Countreymens, I honour, love, and embrace them in the same degree; I was borne in the eighth Climate,* but seeme to bee framed, and constellated unto all; I am no Plant that will not prosper out of a Garden. All places, all ayres make unto me one Countrey; I am in *England*, every where, and under any meridian; I have beene shipwrackt, yet am not enemy with the sea or winds; I can study, play, or sleepe in a tempest. In briefe, I am averse

from nothing, neither Plant, Animal, nor Spirit; my conscience would give me the lie if I should say I absolutely detest or hate any essence but the Devill, or so at least abhorre any thing but that wee might come to composition. If there be any among those common objects of hatred which I can safely say I doe contemne and laugh at, it is that great enemy of reason, vertue and religion, the multitude, that numerous piece of monstrosity, which taken asunder seeme men, and the reasonable creatures of God; but confused together, make but one great beast, & a monstrosity more prodigious than Hydra; it is no breach of Charity to call these fooles; it is the stile all holy Writers have afforded them, set down by *Solomon* in canonicall Scripture, and a point of our faith to beleeve so. Neither in the name of multitude doe I onely include the base and minor sort of people; there is a rabble even amongst the Gentry, a sort of Plebeian heads, whose fancy moves with the same wheele as these; men even in the same Levell with Mechanickes,* though their fortunes doe somewhat guild their infirmities, and their purses compound for their follies. But as in casting account, three or four men together come short of one man placed by himself below them: So neither are a troope of these ignorant Doradoes,* of that true esteeme and value, as many a forlorne person, whose condition doth place him below their feet. Let us speake like Politicians; there is a Nobility without Heraldry, a naturall dignity, whereby one man is ranked with another, another Filed before him, according to the quality of his desert, and preheminence of his good parts. Though the corruption of these times, and the byas of present practice wheele another way, thus it was in the first and primitive Commonwealths, and is yet in the integrity and Cradle of well-ordered polities, till corruption getteth ground, ruder desires labouring after that which wiser considerations contemn, every one having a liberty to amasse & heape up riches, and they a licence or faculty to doe or purchase any thing.

2. This generall and indifferent temper of mine, doth more neerely dispose mee to this noble vertue, that with an easier measure of grace I may obtaine it. It is a happinesse to be borne and framed unto vertue, and to grow up from the seeds of nature, rather than

the inoculation and forced grafts of education; yet if we are directed only by our particular Natures, and regulate our inclinations by no higher rule than that of our reason, we are but Moralists; Divinity will still call us Heathens. Therfore this great worke of charity, must have other motives, ends, and impulsions: I give no almes to satisfie the hunger of my Brother, but to fulfill and accomplish the Will and Command of my God; I draw not my purse for his sake that demands it, but his that hath enjoyned it; I relieve no man upon the Rhetorick of his miseries, nor to content mine owne commiserating disposition, for this is still but morall charity, and an act that oweth more to passion than reason. Hee that relieves another upon the bare suggestion and bowels of pity, doth not this so much for his sake as for his own: for by compassion we make anothers misery our own, & so by relieving them, we relieve our selves also. It is an erroneous course to redresse other mens misfortunes upon the common considerations of mercifull natures, that it may bee one day our own case; for this is a sinister and politick kind of charity, wherby we seem to bespeak the pities of men in the like occasions, buy out of God a faculty to be exempted from it; and truly I have observed that those professed Eleemosynaries,* though in a croud or multitude, doe yet direct and place their petitions on a few and selected persons; there is surely a Physiognomy, which those experienced and Master Mendicants observe, whereby they instantly discover a mercifull aspect, and will single out a face, wherein they spy the signatures and markes of mercy: for there are mystically in our faces certaine characters which carry in them the motto of our Soules, wherein he that cannot read A. B. C. may read our natures. I hold moreover that there is a Phytognomy, or Physiognomy, not onely of men, but of Plants, and Vegetables; and in every one of them, some outward figures which hang as signes or bushes of their inward formes. The finger of God hath set an inscription upon all his workes, not graphicall or composed of Letters, but of their severall formes, constitutions, parts, and operations, which aptly joyned together make one word that doth expresse their natures. By these Letters God cals the Starres by their names, and by this Alphabet

Adam assigned to every creature a name peculiar to its Nature. Now there are besides these Characters in our faces, certaine mysticall lines and figures in our hands, which I dare not call meere dashes, strokes, *a la volee*,* or at randome, because delineated by a pencill, that never workes in vaine; and hereof I take more particular notice, because I carry that in mine owne hand, which I could never read of, or discover in another. *Aristotle*, I confesse, in his acute, and singular booke of Physiognomy, hath made no mention of Chiromancy;* yet I beleeve the *Egyptians*, who were neerer addicted to those abstruse and mysticall sciences, had a knowledge therein, to which those vagabond and counterfeit *Egyptians** did after pretend, and perhaps retained a few corrupted principles, which sometimes might verifie their prognostickes.

It is the common wonder of all men, how among so many millions of faces, there should be none alike; Now contrary, I wonder as much how there should be any; he that shall consider how many thousand severall words have beene carelesly and without study composed out of 24 Letters; withall how many hundred lines there are to be drawn in the fabrick of one man; shall easily finde that this variety is necessary: And it will bee very hard that they should so concur as to make one portract like another. Let a Painter carelesly limn out a Million of faces, and you shall finde them all different; yea let him have his copy before him, yet after all his art there will remaine a sensible distinction; for the pattern or example of every thing is the perfectest in that kind, whereof wee still come short, though wee transcend or goe beyond it, because herein it is wide and agrees not in all points unto its Copy.* I rather wonder how almost all plants being of one colour, yet should bee all different herein, and their severall kinds distinguished in one accident of verte. Nor doth the similitude of creatures disparage the variety of nature, nor any way confound the workes of God. For even in things alike, there is diversitie, and those that doe seeme to accord, doe manifestly disagree. And thus is Man like God, for in the same things that wee resemble him, wee are utterly different from him. There is never any thing so like another, as in all points to concurre; there will ever some reserved difference slip in, to prevent the Identity, without which two severall things would not be alike, but the same, which is impossible.

3. But to returne from Philosophy to Charity, I hold not so narrow a conceit of this vertue, as to conceive that to give almes, is onely to be Charitable, or thinke a piece of Liberality can comprehend the Totall of Charity; Divinity hath wisely divided the acts thereof into many branches, and hath taught us in this narrow way, many pathes unto goodnesse; as many wayes as we may doe good, so many wayes we may bee Charitable; there are infirmities, not onely of body, but of soule, and fortunes, which doe require the mercifull hand of our abilities. I cannot contemn a man for ignorance but behold him with as much pity as I doe *Lazarus.* It is no greater Charity to cloath his body, than apparell the nakednesse of his Soule. It is an honourable object to see the reasons of other men weare our Liveries, and their borrowed understandings doe homage to the bounty of ours. It is the cheapest way of beneficence, and like the naturall charity of the Sunne illuminates another without obscuring it selfe. To be reserved and caitif* in this part of goodnesse, is the sordidest piece of covetousnesse, and more contemptible than pecuniary avarice. To this (as calling my selfe a Scholler) I am obliged by the duty of my condition, I make not therefore my head a grave, but a treasure of knowledge; I intend no Monopoly, but a Community in learning; I study not for my owne sake onely, but for theirs that study not for themselves. I envy no man that knowes more than my selfe, but pity those that know less. I instruct no man as an exercise of my knowledge, or with intent rather to nourish and keepe it alive in mine owne head, than beget and propagate it in his; and in the midst of all my endeavours there is but one thought that dejects me, that my acquired parts must perish with my selfe, nor can bee Legacyed among my honoured Friends. I cannot fall out or contemne a man for an errour, or conceive why a difference in opinion should divide our affection: for controversies, disputes, and argumentations, both in Philosophy and in Divinity, if they meete with discreet and peaceable natures, doe not infringe the Lawes of Charity; in all disputes, so much as there is of passion, so much there is of nothing to the

purpose; for then reason like a bad hound spends upon a false sent, and forsakes the question first started. And this is one reason why controversies are never determined, for though they be amply proposed, they are scarce at all handled, they doe so swell with unnecessary Digressions, and the Parenthesis on the party, is often as large as the maine discourse upon the Subject. The Foundations of Religion are already established, and the principles of Salvation subscribed unto by all; there remaine not many controversies worth a passion, and yet never any disputed without, not onely in Divinity, but in inferiour Arts: What a *βατραχομυομαχία*,* and hot skirmish is betwixt S. and T. in Lucian?* How doe Grammarians hack and slash for the Genitive case in *Iupiter*?[1] How many Synods have been assembled and angerly broke up again about a line in *Propria quæ Maribus?* How doe they break their owne pates to save that of *Priscian*?* *Si foret in terris, rideret Democritus.** Yea, even amongst wiser militants, how many wounds have beene given, and credits stained for the poore victory of an opinion or beggerly conquest of a distinction? Schollers are men of peace, they beare no armes, but their tongues are sharper then *Actius* his razor;[2] their pens carry farther, and give a lowder report than thunder; I had rather stand the shock of a Basilisco,* than the fury of a mercilesse pen. It is not meere zeale to Learning, or Devotion to the Muses, that wiser Princes Patron the Arts, and carry an indulgent aspect unto Schollers; but a desire to have their names eternized by the memory of their writings, and a feare of the revengefull pen of succeeding ages: for these are the men, that when they have played their parts, and had their *exits*, must step out and give the morall of their Scenes, and deliver unto Posterity an Inventory of their vertues and vices. And surely there goes a great deale of conscience to the compiling of an History; there is no reproach to the scandall of a Story; It is such an Authenticke kinde of falsehood that with authority belies our good names to all Nations and Posteritie.

---

1. Whether *Jovis* or *Jupiteris*.

2. That cutt a whetstone in two.

4. There is another offence unto Charity, which no Author hath ever written of, and as few take notice of, and that's the reproach, not of whole professions, mysteries and conditions, but of whole nations, wherein by opprobrious Epithets wee miscall each other, and by an uncharitable Logicke from a disposition in a few conclude a habit in all.

Le mutin Anglois, et le bravache Escossois;
Le bougre Italien, et le fol Francois;
Le poultron Romain, le larron de Gascongne,
*L'Espagnol superbe, et l' Aleman yurongne.**

Saint *Paul** that cals the *Cretians* lyers, doth it but indirectly and upon quotation of their owne Poet. It is as bloody a thought in one way as *Neroes* was in another.* For by a word wee wound a thousand, and at one blow assassine the honour of a Nation. It is as compleate a piece of madnesse to miscall and rave against the times, as thinke to recall men to reason by a fit of passion: *Democritus* that thought to laugh the times into goodnesse, seemes to mee as deeply Hypochondriack, as *Heraclitus* that bewailed them; it moves not my spleene to behold the multitude in their proper humours, that is, in their fits of folly and madnesse, as well understanding that Wisedome is not prophan'd unto the World, and 'tis the priviledge of a few to be vertuous. They that endeavour to abolish vice destroy also vertue, for contraries, though they destroy one another, are yet the life of one another. Thus vertue (abolish vice) is an Idea; againe the communitie of sinne doth not disparage goodnesse; for when vice gaines upon the major part, vertue, in whom it remaines, becomes more excellent, and being lost in some, multiplies its goodnesse in others which remaine untouched, and persists intire in the generall inundation. I can therefore behold vice without a Satyre, content onely with an admonition, or instructive reprehension; for Noble natures, and such as are capable of goodnesse, are railed into vice, but might as easily bee admonished into vertue; and we should be all so farre the Orators of goodnesse, as to protect her from the power of vice, and maintaine the cause of injured truth. No man can justly censure or condemne another, because indeed no man truly knowes another. This I perceive in my selfe, for I am in the darke to all the world, and my nearest friends behold mee but in a cloud; those that know mee but superficially, thinke lesse of me than I doe of my selfe; those of my neere acquaintance thinke more; God, who knowes me truly, knowes that I am nothing; for hee onely beholds me, and all the world, who lookes not on us through a derived ray,* or a trajection of a sensible species,* but beholds the substance without the helpe of accidents, and the formes of things, as wee their operations. Further, no man can judge another, because no man knowes himselfe; for we censure others but as they disagree from that humour which wee fancy laudable in our selves, and commend others but for that wherein they seeme to quadrate* and consent with us. So that in conclusion, all is but that we all condemne, selfe-love. 'Tis the generall complaint of these times, and perhaps of those past, that charity growes cold; which I perceive most verified in those which most doe manifest the fires and flames of zeale; for it is a vertue that best agrees with coldest natures, and such as are complexioned for humility: But how shall we expect charity towards others, when we are uncharitable to our selves? Charity begins at home, is the voyce of the world; yet is every man his greatest enemy, and as it were, his owne executioner. *Non occides*,* is the Commandement of God, yet scarce observed by any man; for I perceive every man is his owne *Atropos*, and lends a hand to cut the thred of his owne dayes. *Cain* was not therefore the first murtherer, but *Adam*, who brought in death; whereof hee beheld the practice onely and example in his owne sonne *Abel*, and saw that verified in the experience of another, which faith could not perswade him in the Theory of himselfe.

5. There is, I thinke, no man that apprehends his owne miseries lesse than my selfe, and no man that so neerely apprehends anothers. I could lose an arme without a teare, and with few groans, mee thinkes, be quartered into pieces; yet can I weepe most seriously at a Play, and receive with a true passion, the counterfeit griefes of those knowne and professed Impostors. It is a barbarous part of inhuman-

ity to adde unto an afflicted parties misery, or endeavour to multiply in any man, a passion, whose single nature is already above his patience; this was the greatest affliction of *Job*, and those oblique expostulations of his friends a deeper injury than the downe-right blowes of the Devill. It is not the teares of our owne eyes onely, but of our friends also, that doe exhaust the current of our sorrowes, which falling asunder into many streames, runs more peaceably within its owne banke, and is contented with a narrower channel. It is an act within the power of charity, to translate a passion out of one breast into another, and to divide a sorrow almost out of it selfe; for an affliction like a dimension* may be so divided, as if not indivisible, at least to become insensible. Now with my friend I desire not to share or participate, but to engrosse his sorrowes, that by making them mine owne, I may more easily discusse them; for in mine owne reason, and within my selfe I can command that, which I cannot entreate without my selfe, and within the circle of another. I have often thought those Noble paires and examples of friendship not so truely Histories of what had beene, as fictions of what should be; but I now perceive nothing therein, but possibilities, nor any thing in the Heroick examples of *Nisus* and *Euryalus*, *Damon* and *Pythias*, *Achilles* and *Patroclus*,* which mee thinkes upon some grounds I could not performe within the narrow compasse of my selfe. That a man should lay down his life for his friend, seemes strange to vulgar affections, and such as confine themselves within that worldly principle, Charity beginnes at home. For mine owne part I could never remember the relations that I hold unto my selfe, nor the respect that I owe unto mine owne nature, in the cause of God, my Country, and my Friends. Next to these three, I doe embrace my selfe; I confesse I doe not observe that order that the Schooles ordaine our affections, to love our Parents, Wives, Children, and then our Friends; for excepting the injunctions of Religion, I doe not finde in my selfe such a necessary and indissoluble Sympathy to all those of my bloud. I hope I doe not breake the fifth Commandement,* if I conceive I may love my friend before the nearest of my bloud, even those to whom I owe the principles of life; I never yet cast a true affection on a Woman, but I have loved my friend as I do vertue, and as I do my soule, my God. From hence me thinkes I doe conceive how God loves man, what happinesse there is in the love of God. Omitting all other there are three most mysticall unions:* Two natures in one person; three persons in one nature; one soule in two bodies. For though indeed they bee really divided, yet are they so united, as they seeme but one, and make rather a duality then two distinct soules.

6. There are wonders in true affection; it is a body of *Ænigmaes*, mysteries and riddles, wherein two so become one, as they both become two; I love my friend before my selfe, and yet me thinkes I do not love him enough; some few months hence my multiplyed affection will make me beleeve I have not loved him at all; when I am from him, I am dead till I bee with him; when I am with him, I am not satisfied, but would still be nearer him; united soules are not satisfied with embraces, but desire each to be truely the other, which being impossible, their desires are infinite, and must proceed without a possibility of satisfaction. Another misery there is in affection: that whom we truely love like our owne selves, wee forget their lookes, nor can our memory retaine the Idea of their faces; and it is no wonder, for they are our selves, and affection makes their lookes our owne. This noble affection fals not on vulgar and common constitutions, but on such as are mark'd for vertue; he that can love his friend with this noble ardour, will in a competent degree affect all. Now, if wee can bring our affections to looke beyond the body, and cast an eye upon the soule, wee have found out the true object, not onely of friendship but charity; and the greatest happinesse that wee can bequeath the soule, is that wherein we all doe place our last felicity: Salvation, which though it bee not in our power to bestow, it is in our charity, and pious invocations to desire, if not procure, and further. I cannot contentedly frame a Prayer for my particular selfe without a catalogue of my friends, nor request a happinesse wherein my sociable disposition doth not desire the fellowship of my neighbour. I never heare the Toll of a passing Bell, though in my mirth, and at a Tavern, without my prayers and best wishes for the departing spirit; I cannot goe to cure the body of my Patient, but I forget

my profession, and call unto God for his soule; I cannot see one say his Prayers, but instead of imitating him, I fall into a supplication for him, who peradventure is no more to mee than a common nature: and if God hath vouchsafed an eare to my supplications, there are surely many happy that never saw me, and enjoy the blessings of mine unknowne devotions. To pray for enemies, that is, for their salvation, is no harsh precept, but the practice of our daily and ordinary devotions. I cannot beleeve the story of the Italian;* our bad wishes and malevolous desires proceed no further than this life; it is the Devill and the uncharitable votes of Hell, that desire our misery in the world to come.

7. To doe no injury, nor take none, was a principle, which to my former yeares, and impatient affections, seemed to containe enough of morality; but my more setled yeares and Christian constitution have fallen upon severer resolutions. I can hold there is no such thing as injury; that if there be, there is no such injury as revenge, and no such revenge as the contempt of an injury; that to hate another, is to maligne himselfe, that the truest way to love another, is to despise our selves. I were unjust unto mine owne conscience, if I should say I am at variance with any thing like my selfe; I finde there are many pieces in this one fabricke of man; and that this frame is raised upon a masse of Antipathies: I am one, mee thinkes, but as the world; wherein notwithstanding there are a swarme of distinct essences, and in them another world of contrarieties; wee carry private and domesticke enemies within, publike and more hostile adversaries without. The Devill that did but buffet Saint *Paul*, playes mee thinkes at sharpes with me: Let mee be nothing if within the compasse of my selfe, I doe not find the battell of *Lepanto*,* passion against reason, reason against faith, faith against the Devill, and my conscience against all. There is another man within mee that's angry with mee, rebukes, commands, and dastards mee.* I have no conscience of Marble to resist the hammer of more heavie offences, nor yet so soft and waxen, as to take the impression of each single peccadillo or scape of infirmity: I am of a strange beliefe, that it is as easie to be forgiven some sinnes, as to commit them. For my original

sinne, I hold it to be washed away in my Baptisme; for my actuall transgressions, I compute and reckon with God, but from my last repentance, Sacrament or generall absolution: And therefore am not terrified with the sinnes and madnesse of my youth. I thanke the goodnesse of God I have no sinnes that want a name; I am not singular in offences, my transgressions are Epidemicall, and from the common breath of our corruption. For there are certaine tempers of body, which matcht with an humorous* depravity of mind, doe hatch and produce viciosities, whose newness and monstrosity of nature admits no name; this was the temper of that Lecher that carnald with a Statua, and the constitution of *Nero* in his Spintrian recreations.* For the heavens are not onely fruitfull in new and unheard of starres, the earth in plants and animals, but mens minds also in villany and vices; now the dulnesse of my reason, and the vulgarity of my disposition, never prompted my invention, nor sollicited my affection unto any of these; yet even those common and *quotidian* infirmities that so necessarily attend me, and doe seeme to bee my very nature, have so dejected me, so broken the estimation that I should have otherwise of my selfe, that I repute my selfe the abjectest piece of mortality; that I detest mine owne nature, and in my retired imaginations cannot withhold my hands from violence on myselfe; Divines prescribe a fit of sorrow to repentance; there goes indignation, anger, contempt, and hatred into mine, passions of a contrary nature, which neither seeme to sute with this action, nor my proper constitution. It is no breach of charity to our selves to be at variance with our vices, nor to abhorre that part of us, which is an enemy to the ground of charity, our God; wherein wee doe but imitate our great selves, the world, whose divided Antipathies and contrary faces doe yet carry a charitable regard unto the whole, by their particular discords preserving the common harmony, and keeping in fetters those powers whose rebellions, once Masters, might bee the ruine of all.

8. I thanke God, amongst those millions of vices I doe inherit and hold from *Adam*, I have escaped one, and that a mortall enemy to charity, the first and father sin, not only of man, but of the devil:

Pride. A vice whose name is comprehended in a Monosyllable, but in its nature circumscribed not with a world; I have escaped it in a condition that can hardly avoid it: those petty acquisitions and reputed perfections that advance and elevate the conceits of other men, adde no feathers unto mine; I have seene a Grammarian towr, and plume himselfe over a single line in *Horace*, and shew more pride in the construction of one Ode, than the Author in the composure of the whole book. For my owne part, besides the *Jargon* and *Patois* of severall Provinces, I understand no less then six Languages; yet I protest I have no higher conceit of my selfe than had our Fathers before the confusion of *Babel*, when there was but one Language in the world, and none to boast himselfe either Linguist or Criticke. I have not onely seene severall Countries, beheld the nature of their climes, the Chorography* of their Provinces, Topography of their Cities, but understand their severall Lawes, Customes and Policies; yet cannot all this perswade the dulnesse of my spirit unto such an opinion of my self, as I behold in nimbler & conceited heads, that never looked a degree beyond their nests. I know the names, and somewhat more, of all the constellations in my Horizon, yet I have seene a prating Mariner that could onely name the Poynters* and the North Starre, out-talke mee, and conceit himselfe a whole Spheare above mee. I know most of the Plants of my Country, and of those about mee; yet me thinkes I do not know so many as when I did but know an hundred, and had scarcely ever Simpled* further than Cheap-side:* for indeed heads of capacity, and such as are not full with a handfull, or easie measure of knowledg, thinke they know nothing, till they know all; which being impossible, they fall upon the opinion of *Socrates*, and onely know they know not any thing. I cannot thinke that *Homer* pin'd away upon the riddle of the Fishermen,* or that *Aristotle*, who understood the uncertainty of knowledge, and so often confessed the reason of man too weake for the workes of nature, did ever drowne himselfe upon the flux and reflux of *Euripus*:* wee doe but learne to day, what our better advanced judgements will unteach us to morrow: and *Aristotle* doth but instruct us as *Plato* did him; that is, to confute himselfe. I have runne through all sorts, yet finde no rest in any; though our first studies & *junior* endeavors may stile us Peripateticks,* Stoicks, or Academicks,* yet I perceive the wisest heads prove at last, almost all Scepticks, and stand like *Janus* in the field of knowledge. I have therefore one common and authentick Philosophy I learned in the Schooles, whereby I discourse and satisfie the reason of other men; another more reserved and drawne from experience whereby I content mine owne. *Solomon* that complained of ignorance in the height of knowledge, hath not onely humbled my conceits, but discouraged my endeavours. There is yet another conceit that hath sometimes made me shut my bookes; which tels mee it is a vanity to waste our dayes in the blind pursuit of knowledge; it is but attending a little longer, and wee shall enjoy that by instinct and infusion which we endeavour at here by labour and inquisition: it is better to sit downe in a modest ignorance, & rest contented with the naturall blessing of our owne reasons, then buy the uncertaine knowledge of this life, with sweat and vexation, which death gives every foole gratis, and is an accessary of our glorification.*

9. I was never yet once married, and commend their resolutions who never marry twice; not that I disallow of second marriage; as neither in all cases of Polygamy, which considering some times and the unequall number of both sexes may bee also necessary. The whole woman was made for man, but the twelfth part of man for woman: man is the whole world and the breath of God, woman the rib onely and crooked piece of man. I could be content that we might procreate like trees, without conjunction, or that there were any way to perpetuate the world without this triviall and vulgar way of coition; It is the foolishest act a wise man commits in all his life, nor is there any thing that will more deject his coold imagination, when hee shall consider what an odde and unworthy piece of folly hee hath committed; I speake not in prejudice, nor am I averse from that sweet sexe, but naturally amorous of all that is beautifull; I can looke a whole day with delight upon a handsome picture, though it be but of an Horse. It is my temper, & I like it the better, to affect all harmony; and sure there is musicke even in beauty, and the silent note

which *Cupid* strikes, farre sweeter than the sound of an instrument. For there is a musicke where-ever there is a harmony, order or proportion; and thus farre we may maintain the musick of the spheares; for those well ordered motions, and regular paces, though they give no sound unto the eare, yet to the understanding they strike a note most full of harmony. Whosoever is harmonically composed delights in harmony; which makes me much distrust the symmetry of those heads which declaime against all Church musicke. For my selfe, not only from my obedience but my particular genius, I doe imbrace it; for even that vulgar and Taverne Musicke, which makes one man merry, another mad, strikes me into a deepe fit of devotion, and a profound contemplation of the first Composer; there is something in it of Divinity more than the eare discovers. It is an Hieroglyphicall and shadowed lesson of the whole world, and the Creatures of God; such a melody to the eare, as the whole world well understood, would afford the understanding. In briefe it is a sensible fit of that Harmony, which intellectually sounds in the eares of God: it unties the ligaments of my frame, takes me to pieces, dilates me out of myself, and by degrees, mee thinkes, resolves me into Heaven. I will not say with *Plato*, the Soule is an Harmony, but harmonicall, and hath its neerest sympathy unto musicke: thus some, whose temper of body agrees, and humours the constitution of their soules, are borne Poets, though indeed all are naturally inclined unto Rhythme. This[3] made *Tacitus* in the very first line of his Story, fall upon a verse; and *Cicero*, the worst of Poets, but declayming[4] for a Poet, fals in the very first sentence upon a perfect Hexameter.[5] I feele not in me those sordid, and unchristian desires of my profession, I doe not secretly implore and wish for Plagues, rejoyce at Famines, revolve Ephemerides, and Almanacks, in expectation of malignant Aspects, fatall conjunctions, and Eclipses: I rejoyce not at unwholsome Springs, nor unseasonable Winters; my Prayers go

3. *Urbem Romam in principio Reges habuere.*

4. *Pro Archia Poeta.*

5. *In qua me non inficior mediocriter esse.*

with the Husbandmans; I desire every thing in its proper season, that neither men nor the times bee out of temper. Let mee be sicke my selfe, if often times the malady of my patient be not a disease unto me. I desire rather to cure his infirmities than my owne necessities. Where I do him no good me thinkes it is scarce honest gaine, though I confess 'tis but the worthy salary of our well intended endeavours. I am not onely ashamed, but heartily sorry, that besides death, there are diseases incurable; yet not for my own sake, or that they be beyond my art, but for the general cause & sake of humanity, whose common cause I apprehend as mine own: And to speak more generally, those three Noble professions which al civil Common wealths doe honour, are raised upon the fall of *Adam*, and are not any exempt from their infirmities; there are not onely diseases incurable in Physicke,* but cases indissoluble in Lawe, Vices incorrigible in Divinity: if general Councells may erre, I doe not see why particular Courts should be infallible: their perfectest rules are raised upon the erroneous reason of Man, and the Lawes of one, doe but condemn the rules of another; as *Aristotle* oft-times the opinions of his predecessours, because, though agreeable to reason, yet were not consonant to his owne rules, and the Logicke of his proper principles. Againe, to speake nothing of the sinne against the Holy Ghost, whose cure not onely, but whose nature is unknowne, I can cure the gout and stone in some, sooner than Divinity, Pride, or Avarice in others. I can cure vices by Physicke, when they remaine incurable by Divinity, and shall obey my pils, when they contemne their precepts. I boast nothing, but plainely say, we all labour against our owne cure, for death is the cure of all diseases. There is no Catholicon,* or universall remedy I know but this, which thogh nauseous to queasier stomachs, yet to prepared appetites is Nectar and a pleasant potion of immortality.

10. For my conversation, it is like the Sunne's with all men; and with a friendly aspect to good and bad. Me thinkes there is no man bad, and the worst, best; that is, while they are kept within the circle of those qualities, wherein they are good: there is no mans minde of such discordant and jarring a temper to which a tuneable disposi-

tion may not strike a harmony. *Magnæ virtutes nec minora vitia** is the posie of the best natures, and may bee inverted on the worst; there are in the most depraved and venemous dispositions, certaine pieces that remaine untoucht; which by an Antiperistasis* become more excellent, or by the excellency of their antipathies are able to preserve themselves from the contagion of their enemy vices, and persist entire beyond the generall corruption. For it is also thus in nature. The greatest Balsames doe lie enveloped in the bodies of the most powerfull Corrosives; I say moreover, and I ground upon experience, that poysons containe within themselves their owne Antidote, and that which preserves them from the venom of themselves; without which they were not deleterious to others onely, but to themselves also. But it is the corruption that I feare within me, not the contagion of commerce without me. 'Tis that unruly regiment within me that will destroy me, 'tis I that doe infect my selfe, the man[6] without a Navell yet lives in me; I feele that originall canker corrode and devoure me, and therefore *Defienda me Dios de mi*, Lord deliver me from my selfe, is a part of my Letany, and the first voyce of my retired imaginations. There is no man alone, because every man is a *Microcosme*, and carries the whole world about him; *Nunquam minus solus quam cum solus,** though it bee the Apophthegme of a wise man, is yet true in the mouth of a foole; for indeed, though in a Wildernesse, a man is never alone, not onely because hee is with himselfe, and his owne thoughts, but because he is with the devill, who ever consorts with our solitude, and is that unruly rebell that musters up those disordered motions, which accompany our sequestred imaginations: And to speake more narrowly, there is no such thing as solitude, nor any thing that can be said to be alone, and by it selfe, but God, who is his owne circle, and can subsist by himselfe; all others, besides their dissimilary and Heterogeneous parts, which in a manner multiply their natures, cannot subsist without the concourse of God, and the society of that hand which doth uphold their natures. In briefe, there can be nothing

6. Adam, whom I conceive to want a navill, because he was not borne of a woman.

truely alone, and by its self, which is not truely one, and such is onely God: All others doe transcend an unity, and so by consequence are many.

11. Now for my life, it is a miracle of thirty yeares, which to relate, were not a History, but a peece of Poetry, and would sound to common eares like a fable; for the world, I count it not an Inne, but an Hospitall, and a place, not to live, but to die in. The world that I regard is my selfe, it is the Microcosme of mine owne frame, that I cast mine eye on; for the other, I use it but like my Globe, and turne it round sometimes for my recreation. Men that look upon my outside, perusing onely my condition, and fortunes, do erre in my altitude; for I am above *Atlas* his shoulders, and though I seeme on earth to stand, on tiptoe in Heaven. The earth is a point not onely in respect of the heavens above us, but of that heavenly and celestiall part within us: that masse of flesh that circumscribes me, limits not my mind: that surface that tells the heavens it hath an end, cannot perswade me I have any; I take my circle to be above three hundred and sixty; though the number of the Arke do measure my body, it comprehendeth not my minde: whilst I study to finde how I am a Microcosme or little world, I finde my selfe something more than the great. There is surely a peece of Divinity in us, something that was before the Elements, and owes no homage unto the Sun. Nature tels me I am the Image of God as well as Scripture; he that understands not thus much, hath not his introduction or first lesson, and is yet to begin the Alphabet of man. Let me not injure the felicity of others, if I say I am as happy as any. I have that in me that can convert poverty into riches, transforme adversity into prosperity. I am more invulnerable than Achilles. Fortune hath not one place to hit me. *Ruat coelum, Fiat voluntas tua** salveth all; so that whatsoever happens, it is but what our daily prayers desire. In briefe, I am content, and what should providence adde more? Surely this is it wee call Happinesse, and this doe I enjoy, with this I am happy in a dreame, and as content to enjoy a happinesse in a fancie as others in a more apparent truth and reality. There is surely a neerer apprehension of any thing that delights us in our dreames, than in our waked senses:

with this I can be a king without a crown, rich without a stiver;* in Heaven though on earth; enjoy my friend and embrace him at a distance, when I cannot behold him; without this I were unhappy, for my awaked judgement discontents me, ever whispering unto me, that I am from my friend; but my friendly dreames in the night requite me, and make me thinke I am within his armes. I thanke God for my happy dreames, as I doe for my good rest, for there is a satisfaction in them unto reasonable desires, and such as can be content with a fit of happinesse; and surely it is not a melancholy conceite to thinke we are all asleepe in this world, and that the conceits of this life are as meare dreames to those of the next, as the Phantasmes of the night, to the conceits of the day. There is an equall delusion in both, and the one doth but seeme to bee the embleme and picture of the other; we are somewhat more than our selves in our sleepes, and the slumber of the body seemes to bee but the waking of the soule. It is the ligation* of sense, but the liberty of reason, and our waking conceptions doe not match the fancies of our sleepes. At my Nativity, my ascendant was the watery signe of *Scorpius*; I was borne in the Planetary houre of *Saturne*, and I think I have a peece of that Leaden Planet* in me. I am no way facetious, nor disposed for the mirth and galliardize* of company; yet in one dreame I can compose a whole Comedy, behold the action, apprehend the jests, and laugh my selfe awake at the conceits thereof; were my memory as faithfull as my reason is then fruitfull, I would never study but in my dreames, and this time also would I chuse for my devotions; but our grosser memories have then so little hold of our abstracted understandings, that they forget the story, and can only relate to our awaked soules, a confused & broken tale of what hath passed. *Aristotle*, who hath written a singular tract of sleepe, hath not me thinkes throughly defined it, nor yet *Galen*, though hee seeme to have corrected it; for those *Noctambuloes* or night-walkers, though in their sleepe, doe yet enjoy the action of their senses; wee must therefore say that there is something in us that is not in the jurisdiction of *Morpheus*; and that those abstracted and ecstaticke soules doe walke about in their owne corps, as spirits in the bodies they assume, wherein they seeme to heare, see and feele, though indeed the organs are destitute of sense, and their natures of those faculties that should informe them. Thus it is observed that men sometimes upon the houre of their departure, doe speake and reason above themselves. For then the soule beginning to bee freed from the ligaments of the body, begins to reason like her selfe, and to discourse in a straine above mortality.

12. We tearme sleepe a death, and yet it is waking that kils us, and destroyes those spirits which are the house of life. Tis indeed a part of life that best expresseth death, for every man truely lives so long as hee acts his nature, or someway makes good the faculties of himselfe: *Themistocles** therefore that slew his Souldier in his sleepe was a mercifull executioner; 'tis a kinde of punishment the mildnesse of no lawes hath invented; I wonder the fancy of *Lucan* and *Seneca** did not discover it. It is that death by which we may be literally said to die daily, a death which *Adam* died before his mortality; a death whereby we live a middle and moderating point betweene life and death; in fine, so like death, I dare not trust it without my prayers, and an halfe adiew unto the world, and take my farewell in a Colloquy with God.

The night is come like to the day,
Depart not thou great God away.
Let not my sinnes, blacke as the night,
Eclipse the lustre of thy light.
Keepe still in my Horizon, for to me,
The Sunne makes not the day, but thee.
Thou whose nature cannot sleepe,
On my temples centry keep;
Guard me 'gainst those watchfull foes,
Whose eyes are open while mine close.
Let no dreames my head infest,
But such as Jacobs *temples blest.*
While I doe rest, my soule advance,
Make my sleepe a holy trance:
That I may, my rest being wrought,
Awake into some holy thought.

And with as active vigour runne
My course, as doth the nimble Sunne.
Sleepe is a death, O make me try,
By sleeping what it is to die.
And down as gently lay my head
Upon my Grave, as now my bed.
How ere I rest, great God let me
Awake againe at last with thee.
And thus assur'd, behold I lie
Securely, whether to wake or die.
These are my drowsie dayes, in vaine
Now I doe wake to sleepe againe.
O come that houre, when I shall never
Sleepe thus againe, but wake for ever!

This is the dormitive[7] I take to bedward; I need no other *Laudanum* than this to make me sleepe; after which I close mine eyes in security, content to take my leave of the Sunne, and sleepe unto the resurrection.

13. The method I should use in distributive justice, I often observe in commutative, and keepe a Geometricall proportion in both, whereby becomming equable to others, I become unjust to my selfe, and supererogate in that common principle* Doe unto others as thou wouldest be done unto thy selfe. I was not borne unto riches, nor is it I thinke my Starre to be wealthy; or if it were, the freedome of my minde, and franknesse of my disposition, were able to contradict and crosse my fates: for to me avarice seemes not so much a vice, as a deplorable piece of madnesse; to conceive our selves Urinals, or bee perswaded that wee are dead, is not so ridiculous, nor so many degrees beyond the power of Hellebore, as this. The opinions of theory and positions of men are not so voyd of reason as their practised conclusions: some have held that Snow is blacke, that the earth moves, that the soule is ayre, fire, water; but all this is Philosophy,

7. The name of an extract wherewith wee use to provoke sleepe.

and there is no *delirium*, if we doe but speculate the folly and indisputable dotage of avarice. To that subterraneous Idoll,* and God of the earth, I doe confesse I am an Atheist; I cannot perswade my selfe to honour that which the world adores; whatsoever vertue its prepared substance may have within my body, it hath no influence nor operation without; I would not entertaine a base designe, or an action that should call mee villaine, for the Indies, and for this onely doe I love and honour my owne soule, and have mee thinkes, two armes too few to embrace my selfe. *Aristotle* is too severe, that will not allow us to be truely liberall without wealth, and the bountiful hand of fortune; if this be true, I must confesse I am charitable onely in my liberall intentions, and bountifull well-wishes. But if the example of the Mite* bee not onely an act of wonder, but an example of the noblest charity, surely poore men may also build Hospitals, and the rich alone have not erected Cathedralls. I have a private method which others observe not: I take the opportunity of my selfe to do good; I borrow occasion of charity from mine owne necessities, and supply the wants of others, when I am most in neede my selfe; when I am reduced to the last tester, I love to divide it with the poore; for it is an honest stratagem to take advantage of our selves, and so to husband the acts of vertue, that where they are defective in one circumstance, they may repay their want, and multiply their goodnesse in another. I have not *Peru** in my desires, but a competence, and abilitie to performe those good workes to which the Almighty hath inclined my nature. Hee is rich, who hath enough to bee charitable, and it is hard to bee so poore, that a noble minde may not finde a way to this piece of goodnesse. *Hee that gives to the poore lendeth to the Lord*; there is more Rhetorick in this one sentence than in a Library of Sermons, and indeed if these sentences were understood by the Reader, with the same Emphasis as they are delivered by the Author, wee needed not those Volumes of instructions, but might bee honest by an Epitome. Upon this motive onely I cannot behold a Begger without relieving his necessities with my purse, or his soule with my prayers; these scenicall and accidentall differences betweene us cannot make mee forget that common and untoucht part of us both;

there is under these *Centoes** and miserable outsides, these mutilate and semi-bodies, a soule of the same alloy with our owne, whose Genealogy is God as well as ours, and in as faire a way unto salvation, as our selves. Statists that labour to contrive a Common-wealth without poverty, take away the object of charity, not understanding only the Common-wealth of a Christian, but forgetting the prophecy of Christ.[8]

14. Now there is another part of charity, which is the Basis and Pillar of this, and that is the love of God, for whom wee love our neighbour: for this I thinke is charity, to love God for himselfe, and our neighbour for God. All that is truely amiable is God, or as it were a divided piece of him, that retaines a reflex or shadow of himselfe. Nor is it strange that wee should place our affection on that which is invisible; all that wee truely love is thus; what wee adore under the affection of our senses, deserves not the honour of so pure a title. Thus wee adore vertue, though to the eyes of sense shee bee invisible. Thus that part of our noble friends that wee love, is not that part that we embrace, but that insensible part that our armes cannot embrace. God being all goodnesse, can love nothing but himselfe; hee loves us but for that part which is as it were himselfe, and the traduction* of his holy Spirit. Let us call to assize* the love of our parents, the affection of our wives and children, and they are all dumb showes, and dreames, without reality, truth, or constancy; for first there is a strong bond of affection betweene us and our parents; yet how easily dissolved? We betake our selves to a woman, forgetting our mother in a wife, the wombe that bare us in that that shall but beare our image. This woman blessing us with children, our affection leaves the levell it held before and sinkes from our bed unto our issue and picture of our posterity, where affection holds no steady mansion. They growing up in yeares either desire our ends, or applying themselves to a woman, take a lawfull way to love another better than our selves. Thus I perceive a man may bee buried alive and behold his grave in his owne issue.

8. The poore ye shall have alwaies with you.

15. I conclude therefore and say, there is no happinesse under (or as *Copernicus*[9] will have it, above) the Sunne, nor any Crambe* in that repeated veritie and burthen of all the wisedom of *Solomon*, *All is vanitie and vexation of spirit*; there is no felicitie in what the world adores. *Aristotle* whilst hee labours to refute the Idea's of *Plato*, fals upon one himselfe: for his *summum bonum*,* is a *Chimæra*, and there is no such thing as his Felicity. That wherein God himselfe is happy, the holy Angels are happy, in whose defect the Devils are unhappy; that dare I call happinesse: whatsoever conduceth unto this, may with an easie Metaphor deserve that name; whatsoever else the world termes happines, is to me a story out of *Pliny*,* an apparition, or neat delusion, wherin there is no more of happinesse than the name. Blesse mee in this life but with the peace of my conscience, command of my affections, the love of thy selfe and my dearest friend, and I shall be happy enough to pity *Cæsar*. These are O Lord the humble desires of my most reasonable ambition and all I dare call happinesse on earth: wherein I set no rule or limit to thy hand or providence. Dispose of me according to the wisedome of thy pleasure. Thy will bee done, though in my owne undoing.

FINIS

9. Who holds the Sunne is the center of the World.

En Sum quod digitis Quinque Levatur onus Propert:

HYDRIOTAPHIA, OR URNE-BURIALL

TO MY WORTHY AND HONOURED FRIEND THOMAS LE GROS OF *CROSTWICK* ESQUIRE

WHEN THE Funerall pyre was out, and the last valediction over, men took a lasting adieu of their interred Friends, little expecting the curiosity of future ages should comment upon their ashes, and having no old experience of the duration of their Reliques, held no opinion of such after considerations.

But who knows the fate of his bones, or how often he is to be buried? who hath the Oracle of his ashes, or whither they are to be scattered? The Reliques of many lie like the ruines of *Pompeys*,[1] in all parts of the earth; And when they arrive at your hands, these may seem to have wandred far, who in a direct[2] and *Meridian* Travell, have but few miles of known Earth between your self and the Pole.

That the bones of *Theseus* should be seen again in *Athens*,[3] was not beyond conjecture, and hopeful expectation; but that these should arise so opportunely to serve your self, was an hit of fate and honour beyond prediction.

We cannot but wish these Urnes might have the effect of Theatrical vessels, and great *Hippodrome* Urnes[4] in *Rome*; to resound the acclamations and honour due unto you. But these are sad and sepulchral Pitchers, which have no joyful voices; silently expressing old mortality, the ruines of forgotten times, and can only speak with life,

1. *Pompeios juvenes Asia, atque Europa, sed ipsum terra tegit Lybies.*
2. Little directly, but Sea between your house and *Greenland*.
3. Brought back by *Cimon*. Plutarch.
4. The great Urnes in the *Hippodrome* at *Rome* conceived to resound the voices of people at their shows.

how long in this corruptible frame, some parts may be uncorrupted; yet able to out-last bones long unborn, and noblest pyle[5] among us.

We present not these as any strange sight or spectacle unknown to your eyes, who have beheld the best of Urnes, and noblest variety of Ashes; Who are your self no slender master of Antiquities, and can daily command the view of so many Imperiall faces; Which raiseth your thoughts unto old things, and consideration of times before you, when even living men were Antiquities; when the living might exceed the dead, and to depart this world could not be properly said, to go unto the greater number.[6] And so run up your thoughts upon the ancient of dayes, the Antiquaries truest object, unto whom the eldest parcels are young, and earth it self an Infant; and without Ægyptian[7] account makes but small noise in thousands.

We were hinted by the occasion, not catched the opportunity to write of old things, or intrude upon the Antiquary. We are coldly drawn unto discourses of Antiquities, who have scarce time before us to comprehend new things, or make out learned Novelties. But seeing they arose as they lay, almost in silence among us, at least in short account suddenly passed over; we were very unwilling they should die again, and be buried twice among us.

Beside, to preserve the living, and make the dead to live, to keep men out of their Urnes, and discourse of humane fragments in them, is not impertinent unto our profession; whose study is life and death, who daily behold examples of mortality, and of all men least need artificial *memento's*, or coffins by our bed side, to minde us of our graves.

'Tis time to observe Occurrences, and let nothing remarkable escape us; The Supinity of elder dayes hath left so much in silence, or time hath so martyred the Records, that the most industrious heads[8] do finde no easie work to erect a new *Britannia*.

---

5. Worthily possessed by that true Gentleman Sir *Horatio Townshend*, my honored Friend.

6. *Abiit ad plures*.

7. Which makes the world so many years old.

8. Wherein Mr *Dugdale* hath excellently well endeavoured, and worthy to be countenanced by ingenuous and noble persons.

'Tis opportune to look back upon old times, and contemplate our Forefathers. Great examples grow thin, and to be fetched from the passed world. Simplicity flies away, and iniquity comes at long strides upon us. We have enough to do to make up our selves from present and passed times, and the whole stage of things scarce serveth for our instruction. A compleat peece of vertue must be made up from the *Centos* of all ages, as all the beauties of *Greece* could make but one handsome *Venus*.

When the bones of King *Arthur* were digged up,[9] the old Race might think, they beheld therein some Originals of themselves; Unto these of our Urnes none here can pretend relation, and can only behold the Reliques of those persons, who in their life giving the Law unto their predecessors, after long obscurity, now lye at their mercies. But remembring the early civility they brought upon these Countreys, and forgetting long passed mischiefs; We mercifully preserve their bones, and pisse not upon their ashes.

In the offer of these Antiquities we drive not at ancient Families, so long out-lasted by them; We are farre from erecting your worth upon the pillars of your Fore-fathers, whose merits you illustrate. We honour your old Virtues, conformable unto times before you, which are the Noblest Armoury. And having long experience of your friendly conversation, void of empty Formality, full of freedome, constant and Generous Honesty, I look upon you as a Gemme of the Old Rock,[10] and must professe my self even to Urne and Ashes,

Your ever faithfull Friend, and Servant,
Thomas Browne.
Norwich
May 1
[1658]

9. In the time of *Henry* the second. *Cambden*.
10. *Adamas de rupe veteri præstantissimus.*

CHAPTER I

IN THE deep discovery of the Subterranean world, a shallow part would satisfie some enquirers; who, if two or three yards were open about the surface, would not care to rake the bowels of *Potosi*,[1] and regions towards the Centre. Nature hath furnished one part of the Earth, and man another. The treasures of time lie high, in Urnes, Coynes, and Monuments, scarce below the roots of some vegetables. Time hath endlesse rarities, and shows of all varieties; which reveals old things in heaven, makes new discoveries in earth, and even earth it self a discovery. That great Antiquity *America* lay buried for thousands of years; and a large part of the earth is still in the Urne unto us.

Though if *Adam* were made out of an extract of the Earth, all parts might challenge a restitution,* yet few have returned their bones farre lower than they might receive them;* not affecting the graves of Giants, under hilly and heavy coverings, but content with lesse than their owne depth, have wished their bones might lie soft, and the earth be light upon them; Even such as hope to rise again, would not be content with centrall interrment,* or so desperately to place their reliques as to lie beyond discovery, and in no way to be seen again; which happy contrivance hath made communication with our forefathers, and left unto our view some parts, which they never beheld themselves.*

Though earth hath engrossed the name yet water hath proved the

1. The rich mountain of *Peru*.

smartest grave; which in forty dayes swallowed almost mankinde, and the living creation; Fishes not wholly escaping, except the Salt Ocean were handsomely contempered by admixture of the fresh Element.

Many have taken voluminous pains to determine the state of the soul upon disunion; but men have been most phantasticall in the singular contrivances of their corporall dissolution; whilest the sobrest Nations have rested in two wayes, of simple inhumation and burning.

That carnall interment or burying, was of the elder date, the old examples of *Abraham* and the Patriarchs are sufficient to illustrate; And were without competition, if it could be made out, that *Adam* was buried near *Damascus,* or Mount *Calvary*, according to some Tradition. God himself, that buried but one, was pleased to make choice of this way, collectible from Scripture-expression, and the hot contest between Satan and the Arch-Angel, about discovering the body of *Moses*.* But the practice of Burning was also of great Antiquity, and of no slender extent. For (not to derive the same from *Hercules*) noble descriptions there are hereof in the Grecian Funerals of *Homer,* In the formall Obsequies of *Patroclus,* and *Achilles*; and somewhat elder in the *Theban* warre, and solemn combustion of *Meneceus*, and *Archemorus*, contemporary unto *Jair* the Eighth Judge of *Israel.* Confirmable also among the *Trojans*, from the Funerall Pyre of *Hector*, burnt before the gates of *Troy*, And the burning of *Penthisilea* the *Amazonean Queen:*[2] and long continuance of that practice, in the inward Countries of *Asia*; while as low as* the Reign of *Julian*, we finde that the King of *Chionia*[3] burnt the body of his Son, and interred the ashes in a silver Urne.

The same practice extended also farre West,[4] and besides *Herulians, Getes,* and *Thracians*, was in use with most of the *Celtæ, Sarmatians, Germans, Gauls, Danes, Swedes, Norwegians*; not to omit

2. Q. Calaber lib. 1.

3. Gumbrates King of *Chionia,* a countrey near Persia. Ammianus Marcellinus.

4. Arnoldi Montani *not. in Cæs. Commentar.* L. Gyraldus. Kirkmannus.

some use thereof among *Carthaginians* and *Americans*: Of greater Antiquity among the *Romans* than most opinion, or *Pliny* seems to allow. For (beside the old Table Laws of burning[5] or burying within the City, of making the Funerall fire with plained wood, or quenching the fire with wine) *Manlius* the Consul burnt the body of his Son: *Numa* by speciall clause of his Will, was not burnt but buried; And *Remus* was solemnly burned, according to the description of *Ovid*.[6]

Cornelius Sylla was not the first whose body was burned in *Rome*, but of the *Cornelian* Family, which being indifferently, not frequently used before, from that time spread, and became the prevalent practice. Not totally pursued in the highest runne of Cremation;* For when even Crows were funerally burnt, *Poppæa* the Wife of *Nero* found a peculiar grave enterment. Now as all customes were founded upon some bottome of Reason, so there wanted not grounds for this; according to severall apprehensions of the most rationall dissolution. Some being of the opinion of *Thales*, that water was the originall of all things, thought it most equall to submit unto the principle of putrefaction, and conclude in a moist relentment.* Others conceived it most natural to end in fire, as due unto the master principle in the composition, according to the doctrine of *Heraclitus*. And therefore heaped up large piles, more actively to waft them toward that Element, whereby they also declined a visible degeneration into worms, and left a lasting parcell of their composition.

Some apprehended a purifying virtue in fire, refining the grosser commixture, and firing out the Æthereall particles so deeply immersed in it. And such as by tradition or rationall conjecture held any hint of the finall pyre of all things; or that this Element at last must be too hard for all the rest; might conceive most naturally of the fiery dissolution. Others pretending no natural grounds, politickly

5. 12. *Tabul. part. 1. de jure sacro. Hominem mortuum in urbe ne sepelito, neve urito. tom. 2. Rogum ascia ne polito. tom. 4. Item* Vigeneri *Annotat. in Livium, &* Alex. ab Alex. *cum* Tiraquello. Roscinus *cum* Dempstero.

6. *Ultima plorato subdita flamma rogo. De Fast., lib.* 4 *cum* Car. Neapol. *anaptyxi*.

declined the malice of enemies upon their buried bodies. Which consideration led *Sylla* unto this practise; who having thus served the body of *Marius*, could not but fear a retaliation upon his own, entertained after in the Civill wars, and revengeful contentions of *Rome*.

But as many Nations embraced, and many left it indifferent, so others too much affected, or strictly declined this practice. The *Indian Brachmans* seemed too great friends unto fire, who burnt themselves alive, and thought it the noblest way to end their dayes in fire; according to the expression of the Indian, burning himself at *Athens*,[7] in his last words upon the pyre unto the amazed spectators, *Thus I make my selfe Immortall*.

But the *Chaldeans*, the great Idolaters of fire, abhorred the burning of their carcasses, as a pollution of that Deity. The *Persian Magi* declined it upon the like scruple, and being only sollicitous about their bones, exposed their flesh to the prey of Birds and Dogges. And the *Persees* now in *India*, which expose their bodies unto Vultures, and endure not so much as *feretra* or Beers of Wood, the proper Fuell of fire, are led on with such niceties. But whether the ancient *Germans* who burned their dead, held any such fear to pollute their Deity of *Herthus*, or the earth, we have no Authentick conjecture.

The Ægyptians were afraid of fire, not as a Deity, but a devouring Element, mercilesly consuming their bodies, and leaving too little of them; and therefore by precious Embalments, depositure in dry earths, or handsome inclosure in glasses, contrived the notablest wayes of integrall conservation. And from such Ægyptian scruples imbibed by *Pythagoras*, it may be conjectured that *Numa* and the Pythagoricall Sect first waved the fiery solution.*

The *Scythians* who swore by winde and sword, that is, by life and death, were so farre from burning their bodies, that they declined all interrment, and made their graves in the ayr: And the *Ichthyophagi* or fish-eating Nations about Ægypt, affected* the Sea for their grave: Thereby declining visible corruption, and restoring the debt of their bodies. Whereas the old Heroes in *Homer* dread nothing more than

7. And therefore the Inscription of his Tomb was made accordingly. Nic. Damasc.

water or drowning; probably upon the old opinion of the fiery substance of the soul, only extinguishable by that Element; And therefore the Poet emphatically implieth the totall destruction in this kinde of death, which happened to *Ajax Oileus.*[8]

The old *Balearians*[9] had a peculiar mode, for they used great Urnes and much wood, but no fire in their burials, while they bruised the flesh and bones of the dead, crowded them into Urnes, and laid heapes of wood upon them. And the *Chinois*[10] without cremation or urnall interrment of their bodies, make use of trees and much burning, while they plant a Pine-tree by their grave, and burn great numbers of printed draughts of slaves and horses over it, civilly content with their companies in effigie, which barbarous Nations exact unto reality.

Christians abhorred this way of obsequies, and though they stickt not to give their bodies to be burnt in their lives, detested that mode after death; affecting rather a depositure than absumption,* and properly submitting unto the sentence of God, to return not unto ashes but unto dust againe, conformable unto the practice of the Patriarchs, the interrment of our Saviour, of *Peter*, *Paul*, and the ancient Martyrs. And so farre at last declining promiscuous enterrment with Pagans, that some have suffered Ecclesiastical censures,[11] for making no scruple thereof.

The *Musselman* beleevers will never admit this fiery resolution. For they hold a present trial from their black and white Angels in the grave; which they must have made so hollow, that they may rise upon their knees.

The Jewish Nation, though they entertained the old way of inhumation, yet sometimes admitted this practice. For the men of *Jabesh* burnt the body of *Saul.** And by no prohibited practice, to avoid contagion or pollution, in time of pestilence, burnt the bodies of their

---

8. Which Magius reads *ἐξαπόλωλε.*

9. Diodorus Siculus.

10. Ramusius in *Navigat.*

11. Martialis the Bishop, Cyprian.

friends.[12] And when they burnt not their dead bodies, yet sometimes used great burnings neare and about them, deducible from the expressions concerning *Jehoram*, *Sedechias*, and the sumptuous pyre of *Asa*:* And were so little averse from Pagan burning, that the Jews lamenting the death of *Cæsar* their friend, and revenger on *Pompey*, frequented the place where his body was burnt for many nights together.[13] And as they raised noble Monuments and *Mausolæums* for their own Nation,[14] so they were not scrupulous in* erecting some for others, according to the practice of *Daniel*, who left that lasting sepulchrall pyle in *Echbatana*, for the *Medean* and *Persian* Kings.[15]

But even in times of subjection and hottest use,* they conformed not unto the *Romane* practice of burning; whereby the Prophecy was secured concerning the body of Christ, that it should not see corruption, or a bone should not be broken; which we beleeve was also providentially prevented, from the Souldiers spear and nails that past by the little bones both in his hands and feet: Nor of ordinary contrivance, that it should not corrupt on the Crosse, according to the Laws of *Romane* Crucifixion, or an hair of his head perish, though observable in Jewish customes, to cut the hairs of Malefactors.

Nor in their long co-habitation with Ægyptians, crept into a custome of their exact embalming, wherein deeply slashing the muscles, and taking out the brains and entrails, they had broken the subject of so entire a Resurrection, nor fully answered the types of *Enoch, Eliah*, or *Jonah*,* which yet to prevent or restore, was of equall facility unto that rising power, able to break the fasciations* and bands of death, to get clear out of the Cere-cloth, and an hundred pounds of oyntment, and out of the Sepulchre before the stone was rolled from it.

But though they embraced not this practice of burning, yet entertained they many ceremonies agreeable unto *Greeke* and *Romane* obsequies. And he that observeth their funerall Feasts, their Lamen-

12. Amos 6. 10.

13. Sueton. *in vita Jul. Cæs.*

14. As that magnificent Monument erected by Simon. 1 Macc. 13.

15. *κατασκεύασμα θαυμασίως πεποιημένον*, whereof a Jewish Priest had alwayes the custody unto *Josephus* his dayes. Jos. Lib. 10. Antiq.

tations at the grave, their musick, and weeping mourners; how they closed the eyes of their friends, how they washed, anointed, and kissed the dead; may easily conclude these were not meere Pagan-Civilities. But whether that mournfull burthen, and treble calling out after *Absalom*,[16] had any reference unto the last conclamation, and triple valediction, used by other Nations, we hold but a wavering conjecture.

Civilians make sepulture but of the Law of Nations, others doe naturally found it and discover it also in animals.* They that are so thick skinned as still to credit the story of the *Phœnix*, may say something for animall burning: More serious conjectures finde some examples of sepulture in Elephants, Cranes, the Sepulchrall Cells of Pismires and practice of Bees; which civill society carrieth out their dead, and hath exequies, if not interrments.

CHAPTER II

The Solemnities, Ceremonies, Rites of their Cremation or enterrment, so solemnly delivered by Authours, we shall not disparage our Reader to repeat. Only the last and lasting part in their Urns, collected bones and Ashes, we cannot wholly omit, or decline that Subject, which occasion lately presented, in some discovered among us.

In a Field of old *Walsingham*, not many moneths past, were digged up between fourty and fifty Urnes, deposited in a dry and sandy soile, not a yard deep, nor farre from one another: Not all strictly of one figure, but most answering these described: Some containing two pounds of bones, distinguishable in skulls, ribs, jawes, thigh-bones, and teeth, with fresh impressions of their combustion. Besides the extraneous substances, like peeces of small boxes, or combes handsomely wrought, handles of small brasse instruments, brazen nippers, and in one some kinde of *Opale*.[17]

16. *O Absolom, Absolom, Absolom*. 2 Sam. 18.
17. In one sent me by my worthy friend Dr *Thomas Witherley* of *Walsingham*.

Near the same plot of ground, for about six yards compasse were digged up coals and incinerated substances, which begat conjecture that this was the *Ustrina** or place of burning their bodies, or some sacrificing place unto the *Manes,** which was properly below the surface of the ground, as the *Aræ** and Altars unto the gods and *Heroes* above it.

That these were the Urnes of *Romanes* from the common custome and place where they were found, is no obscure conjecture, not farre from a *Romane* Garrison, and but five Miles from *Brancaster,* set down by ancient Record under the name of *Brannodunum.* And where the adjoyning Towne, containing seven Parishes, in no very different sound, but Saxon Termination, still retains the Name of *Burnham*, which being an early station, it is not improbable the neighbour parts were filled with habitations, either of *Romanes* themselves, or *Brittains Romanised*, which observed the *Romane* customes.

Nor is it improbable that the *Romanes* early possessed this Countrey; for though we meet not with such strict particulars of these parts, before the new Institution of *Constantine,* and military charge of the Count of the *Saxon* shore, and that about the *Saxon* Invasions, the *Dalmatian* Horsemen were in the Garrison of *Brancaster*: Yet in the time of *Claudius, Vespasian*, and *Severus*, we finde no lesse than three Legions dispersed through the Province of *Brittain.*[18] And as high as the Reign of *Claudius* a great overthrow was given unto the *Iceni,* by the *Romane* Lieutenant *Ostorius.* Not long after the Countrey was so molested, that in hope of a better state, *Prasutagus* bequeathed his Kingdome unto *Nero* and his Daughters; and *Boadicea* his Queen fought the last decisive Battle with *Paulinus.* After which time and Conquest of *Agricola* the Lieutenant of *Vespasian*, probable it is they wholly possessed this Countrey, ordering it into Garrisons or Habitations, best suitable with their securities. And so some *Romane* Habitations, not improbable in these parts, as high as the time of *Vespasian*, where the *Saxons* after seated,

18. In Onuphrius.

in whose thin-fill'd Mappes we yet finde the Name of *Walsingham*. Now if the *Iceni* were but *Gammadims, Anconians,* or men that lived in an Angle wedge or Elbow of *Brittain,* according to the Originall Etymologie, this countrey will challenge the Emphaticall appellation,* as most properly making the Elbow or Iken of *Icenia*.

That *Britain* was notably populous is undeniable, from that expression of *Cæsar.*[19] That the *Romans* themselves were early in no small Numbers, Seventy Thousand with their associats slain by *Boadicea*, affords a sure account. And though many *Roman* habitations are now unknowne, yet some by old works, Rampiers, Coynes, and Urnes doe testifie their Possessions. Some Urnes have been found at *Castor,* some also about *Southcreake*, and not many years past, no lesse than ten in a Field at *Buxton*,[20] not near any recorded Garison. Nor is it strange to finde *Romane* Coynes of Copper and Silver among us; of *Vespasian, Trajan, Adrian, Commodus, Antoninus, Severus,* &c. But the greater number of *Dioclesian, Constantine, Constans, Valens,* with many of *Victorinus, Posthumius, Tetricus,* and the thirty Tyrants in the Reigne of *Gallienus*; and some as high as *Adrianus* have been found about *Thetford*, or *Sitomagus*, mentioned in the itinerary of *Antoninus*, as the way from *Venta* or *Castor* unto *London*.[21] But the most frequent discovery is made at the two *Casters* by *Norwich* and *Yarmouth*,[22] at *Burghcastle* and *Brancaster.*[23]

---

19. *Hominum infinita multitudo est, creberrimaque ædificia fere Gallicis consimilia. Cæs. de bello Gal.* 1.5.

20. In the ground of my worthy Friend *Rob. Jegon* Esq. wherein contained were preserved by the most worthy Sir William Paston, Bt.

21. From *Castor to Thetford* the Romans accounted thirty-two miles, and from thence observed not our common road to *London*, but passed by *Combretonium* ad *Ansam, Canonium, Cæsaromagus,* &c. by *Bretemham, Coggeshall, Chelmeford, Burntwood,* &c.

22. Most at *Caster* by *Yarmouth*, found in a place called *East-bloudy-burgh furlong*, belonging to Mr *Thomas Wood*, a person of civility, industry and knowledge in this way, who hath made observation of remarkable things about him, and from whom we have received divers Silver and Copper Coynes.

23. Belonging to that Noble Gentleman, and true example of worth Sir *Ralph Hare* Baronet, my honoured Friend.

Besides the *Norman*, *Saxon* and *Danish* peeces of *Cuthred*, *Canutus*, *William*, *Matilda*,[24] and others, som Brittish Coynes of gold have been dispersedly found; And no small number of silver peeces near[25] *Norwich*; with a rude head upon the obverse, and an ill formed horse on the reverse, with Inscriptions *Ic. Duro. T.* whether implying *Iceni*, *Durotriges*, *Tascia*, or *Trinobantes*, we leave to higher conjecture.* Vulgar Chronology will have *Norwich* Castle as old as *Julius Cæsar*; but his distance from these parts, and its *Gothick* form of structure, abridgeth such Antiquity.* The *British* Coyns afford conjecture of early habitation in these parts, though the City of *Norwich* arose from the ruines of *Venta*, and though perhaps not without some habitation before, was enlarged, builded, and nominated by the *Saxons*. In what bulk or populosity it stood in the old East-angle Monarchy, tradition and history are silent. Considerable it was in the *Danish* Eruptions, when *Sueno* burnt *Thetford* and *Norwich*,[26] and *Ulfketel* the Governour thereof was able to make some resistance, and after endeavoured to burn the *Danish* Navy.

How the *Romanes* left so many Coynes in Countreys of their Conquests, seems of hard resolution, except we consider how they buried them under ground, when upon barbarous invasions they were fain to desert their habitations in most part of their Empire; and the strictnesse of their laws forbidding to transfer them to any other uses; Wherein the *Spartans*[27] were singular, who to make their Copper money uselesse, contempered it with vinegar. That the *Brittains* left any, some wonder; since their money was iron, and Iron rings before *Cæsar*; and those of after stamp by permission, and but small in bulk and bignesse. That so few of the *Saxons* remain, because overcome by succeeding Conquerours upon the place, their Coynes by degrees passed into other stamps, and the marks of after ages.

---

24. A peece of *Maud* the Empresse said to be found in *Buckenham* Castle with this Inscription, *Elle n'a elle*.

25. At *Thorpe*.

26. *Brampton Abbas Jorvallensis*.

27. Plut. *in Vita Lycurg*.

Than the time of these Urnes deposited, or precise Antiquity of these Reliques, nothing of more uncertainty. For since the Lieutenant of *Claudius* seems to have made the first progresse into these parts, since *Boadicea* was overthrown by the Forces of *Nero*, and *Agricola* put a full end to these Conquests; it is not probable the Countrey was fully garrison'd or planted before; and therefore however these Urnes might be of later date, not likely of higher Antiquity.

And the succeeding Emperours desisted not from their Conquests in these and other parts; as testified by history and medall inscription yet extant. The Province of *Brittain* is so divided a distance from *Rome*, beholding the faces of many Imperiall persons, and in large account no fewer than *Cæsar, Claudius, Britannicus, Vespasian, Titus, Adrian, Severus, Commodus, Geta*, and *Caracalla*.

A great obscurity herein, because no medall or Emperours Coyne enclosed, which might denote the date of their enterrments; observable in many Urnes, and found in those of *Spittle* Fields by *London*,[28] which contained the Coynes of *Claudius, Vespasian, Commodus, Antoninus*, attended with Lacrymatories,* Lamps, Bottles of Liquor, and other appurtenances of affectionate superstition, which in these rurall interrements were wanting.

Some uncertainty there is from the period or term of burning, or the cessation of that practise. *Macrobius* affirmeth it was disused in his dayes. But most agree, though without authentick record, that it ceased with the *Antonini*. Most safely to be understood, after the Reigne of those Emperours which assumed the name of *Antoninus*, extending unto *Heliogabalus*. Not strictly after *Marcus*; For about fifty years later we finde the magnificent burning, and consecration of *Severus*; and if we so fix this period or cessation, these Urnes will challenge above* thirteen hundred years.

But whether this practise was onely then left by Emperours and great persons, or generally about *Rome*, and not in other Provinces, we hold no authentick account. For after *Tertullian*, in the dayes of *Minucius* it was obviously objected upon Christians,* that they

28. Stowe's *Survey of London*.

condemned the practise of burning.[29] And we finde a passage in *Sidonius*,[30] which asserteth that practise in *France* unto a lower account. And perhaps not fully disused till Christianity fully established, which gave the finall extinction to these sepulchrall Bonefires.

Whether they were the bones of men or women or children, no authentick decision from ancient custome in distinct places of buriall. Although not improbably conjectured, that the double Sepulture or burying place of *Abraham*,[31] had in it such intension. But from exility* of bones, thinnesse of skulls, smallnesse of teeth, ribbes, and thigh-bones; not improbable that many thereof were persons of *minor* age, or women. Confirmable also from things contained in them: In most were found substances resembling Combes, Plates like Boxes, fastened with Iron pins, and handsomely overwrought like the necks or Bridges of Musicall Instruments, long brasse plates overwrought like the handles of neat implements, brazen nippers to pull away hair, and in one a kinde of *Opale* yet maintaining a blewish colour.

Now that they accustomed to burn or bury with them things wherein they excelled, delighted, or which were dear unto them, either as farewells unto all pleasure, or vain apprehension that they might use them in the other world, is testified by all Antiquity. Observable from the Gemme or Berill Ring upon the finger of *Cynthia*, the Mistresse of *Propertius*, when after her Funerall Pyre her Ghost appeared unto him. And notably illustrated from the Contents of that *Romane* Urne preserved by Cardinall *Farnese*,[32] wherein besides great number of Gemmes with heads of Gods and Goddesses, were found an Ape of *Agath*, a Grashopper, an Elephant of Ambre, a Crystall Ball, three glasses, two Spoones, and six Nuts of Crystall. And beyond the content of Urnes, in the Monument of *Childerick* the first,[33] and fourth King from *Pharamond*, casually discovered three years past at *Tournay*, restoring unto the world much gold

29. *Execrantur rogos, & damnant ignium sepulturam*. Min. in Oct.
30. Sidon. Apollinaris.
31. *Det mihi speluncam duplicem*. Gen. 23.
32. *Vigeneri Annot. in* 4. Liv.
33. Chifflet in *Anast. Childer.*

richly adorning his Sword, two hundred Rubies, many hundred Imperial Coyns, three hundred golden Bees, the bones and horseshoe of his horse enterred with him, according to the barbarous magnificence of those dayes in their sepulchral Obsequies. Although if we steer by the conjecture of many and Septuagint* expression; some trace thereof may be found even with the ancient Hebrews, not only from the Sepulcrall treasure of *David*, but the circumcision knives which *Josuah* also buried.

Some men considering the contents of these Urnes, lasting peeces and toyes included in them, and the custome of burning with many other Nations, might somewhat doubt whether all Urnes found among us were properly *Romane* Reliques, or some not belonging unto our *Brittish, Saxon*, or *Danish* Forefathers.

In the form of Buriall among the ancient *Brittains*, the large Discourses of *Cæsar, Tacitus*, and *Strabo* are silent: For the discovery whereof, with other particulars, we much deplore the losse of that Letter which *Cicero* expected or received from his Brother *Quintus*, as a resolution of *Brittish* customes; or the account which might have been made by *Scribonius Largus*, the Physician accompanying the Emperour *Claudius*, who might have also discovered that frugall Bit* of the Old *Brittains*,[34] which in the bignesse of a Bean could satisfie their thirst and hunger.

But that the *Druids* and ruling Priests used to burn and bury, is expressed by *Pomponius*; That *Bellinus* the Brother of *Brennus* and King of *Brittains* was burnt, is acknowledged by *Polydorus*.[35] That they held that practise in *Gallia*, *Cæsar* expresly delivereth. Whether the *Brittains* (probably descended from them, of like Religion, Language and Manners) did not sometimes make use of burning; or whether at least such as were after civilized unto the *Romane* life and manners, conformed not unto this practise, we have not historicall assertion or deniall. But since from the account of *Tacitus* the *Romanes*

34. *Dionis excerpta per Xiphilin. in Severo.*

35. As also by Amandus Zierexensis in *Historia*, and Pineda in his *Universa historia*, Spanish.

early wrought so much civility upon the British stock, that they brought them to build Temples, to wear the Gowne, and study the *Romane* Laws and language, that they conformed also unto their religious rites and customes in burials, seems no improbable conjecture.

That burning the dead was used in *Sarmatia*, is affirmed by *Gaguinus*, that the *Sueons* and *Gothlanders* used to burne their Princes and great persons, is delivered by *Saxo* and *Olaus*; that this was the old *Germane* practise, is also asserted by *Tacitus*. And though we are bare in historicall particulars of such obsequies in this Island, or that the *Saxons, Jutes*, and *Angles* burnt their dead, yet came they from parts where 'twas of ancient practise; the *Germanes* using it, from whom they were descended. And even in *Jutland* and *Sleswick* in *Anglia Cymbrica*, Urnes with bones were found not many years before us.

But the *Danish* and Northern Nations have raised an Æra* or point of compute from their Custome of burning their dead:[36] Some deriving it from *Unguinus*, some from *Frotho* the great; who ordained by Law, that Princes and Chief Commanders should be committed unto the fire, though the common sort had the common grave enterrment. So *Starkatterus* that old *Heroe* was burnt, and *Ringo* royally burnt the body of *Harald* the King slain by him.

What time this custome generally expired in that Nation, we discern no assured period; whether it ceased before Christianity, or upon their Conversion, by *Ansgarius* the Gaul in the time of *Ludovicus Pius* the Sonne of *Charles* the great, according to good computes; or whether it might not be used by some persons, while for a hundred and eighty years Paganisme and Christianity were promiscuously embraced among them, there is no assured conclusion. About which times the *Danes* were busie in *England*, and particularly infested this Countrey: Where many Castles and strong holds were built by them, or against them, and great number of names and Families still derived from them. But since this custome was probably disused before their Invasion or Conquest, and the *Romanes* confessedly prac-

36. Roisold, Brendetiide. Ild tyde.

tised the same, since their possession of this Island, the most assured account will fall upon the *Romanes*, or *Brittains Romanized*.

However, certain it is, that Urnes conceived of no *Romane* Originall, are often digged up both in *Norway*, and *Denmark*, handsomely described, and graphically represented by the Learned Physician *Wormius*,[37] And in some parts of *Denmark* in no ordinary number, as stands delivered by Authours exactly describing those Countreys.[38] And they contained not only bones, but many other substances in them, as Knives, peeces of Iron, Brasse and Wood, and one of *Norwaye* a brasse guilded Jewes-harp.

Nor were they confused or carelesse in disposing the noblest sort, while they placed large stones in circle about the Urnes, or bodies which they interred: Somewhat answerable unto the Monument of *Rollrich* stones in England,[39] or sepulcrall Monument probably erected by *Rollo*, who after conquered *Normandy*. Where 'tis not improbable somewhat might be discovered. Mean while to what Nation or person belonged that large Urne found at *Ashburie*,[40] containing mighty bones, and a Buckler; What those large Urnes found at little *Massingham*,[41] or why the *Anglesea* Urnes are placed with their mouths downward, remains yet undiscovered.

CHAPTER III

Playstered and whited Sepulchres were anciently affected in cadaverous and corruptive Burials; And the rigid Jews were wont to garnish the Sepulchres of the righteous;[42] *Ulysses* in Hecuba*[43] cared not how meanly he lived, so he might finde a noble Tomb after

37. *Olai Wormii monumenta & Antiquitat. Dan.*
38. Adolphus Cyprius in *Annal. Sleswic., urnis adeo abundabat collis*; &c.
39. In Oxford-shire. Cambden.
40. In Cheshire, Twinus *de rebus Albionicis*.
41. In Norfolk, Hollingshead.
42. Mat. 23.
43. Euripides.

death. Great Persons affected great Monuments, And the fair and larger Urnes contained no vulgar ashes, which makes that disparity in those which time discovereth among us. The present Urnes were not of one capacity, the largest containing above a gallon, Some not much above half that measure; nor all of one figure, wherein there is no strict conformity, in the same or different Countreys; Observable from those represented by *Casalius, Bosio*, and others, though all found in *Italy*: While many have handles, ears, and long necks, but most imitate a circular figure, in a sphericall and round composure; whether from any mystery, best duration or capacity, were but a conjecture. But the common form with necks was a proper figure, making our last bed like our first;* nor much unlike the Urnes of our Nativity, while we lay in the nether part of the Earth,[44] and inward vault of our Microcosme.* Many Urnes are red, these but of a black colour, somewhat smooth, and dully sounding, which begat some doubt, whether they were burnt, or only baked in Oven or Sunne: According to the ancient way, in many bricks, tiles, pots, and testaceous* works; and as the word *testa** is properly to be taken, when occurring without addition: And chiefly intended by *Pliny*, when he commendeth bricks and tiles of two years old, and to make them in the spring. Nor only these concealed peeces, but the open magnificence of Antiquity, ran much in the Artifice of Clay. Hereof the house of *Mausolus* was built, thus old *Jupiter* stood in the Capitoll, and the *Statua* of *Hercules* made in the Reign of *Tarquinius Priscus*, was extant in *Plinies* dayes. And such as declined burning or Funerall Urnes, affected Coffins of Clay, according to the mode of *Pythagoras*, and way preferred by *Varro*. But the spirit of great ones was above these circumscriptions, affecting copper, silver, gold, and *Porphyrie* Urnes, wherein *Severus* lay, after a serious view and sentence on that which should contain him.[45] Some of these Urnes were thought to have been silvered over, from sparklings in several pots,

44. Psa. 63.

45. *χωρήσεις τον ἄνθρωπον, ὃν ἥ οἰκουμένη οὐκ ἐχώρησεν* Dion.

with small Tinsell parcels; uncertain whether from the earth, or the first mixture in them.

Among these Urnes we could obtain no good account of their coverings; Only one seemed arched over with some kinde of brick-work. Of those found at *Buxton* some were covered with flints, some in other parts with tiles, those at *Yarmouth Caster* were closed with *Romane* bricks. And some have proper earthen covers adapted and fitted to them. But in the *Homericall* Urne of *Patroclus*, whatever was the solid Tegument,* we finde the immediate covering to be a purple peece of silk: And such as had no covers might have the earth closely pressed into them, after which disposure were probably some of these, wherein we found the bones and ashes half mortered unto the sand and sides of the Urne; and some long roots of Quich, or Dogs-grass wreathed about the bones.

No Lamps, included Liquors, Lachrymatories, or Tear-bottles attended these rurall Urnes, either as sacred unto the *Manes*, or passionate expressions of their surviving friends. While with rich flames and hired tears they solemnized their Obsequies, and in the most lamented Monuments made one part of their Inscriptions.[46] Some finde sepulchrall Vessels containing liquors, which time hath incrassated* into gellies. For beside these Lachrymatories, notable Lamps with Vessels of Oyles and Aromaticall Liquors attended noble Ossuaries. And some yet retaining a Vinosity[47] and spirit in them, which if any have tasted they have farre exceeded the Palats of Antiquity.* Liquors not to be computed by years of annuall Magistrates, but by great conjunctions and the fatall periods of Kingdomes.[48] The draughts of Consulary date* were but crude unto these, and *Opimian*[49] Wine* but in the must unto them.

In sundry Graves and Sepulchres, we meet with Rings, Coynes,

46. *Cum lacrymis posuere.*

47. Lazius.

48. About five hundred years. Plato.

49. *Vinum opimianum annorum centum.* Petron.

and Chalices; Ancient frugality was so severe, that they allowed no gold to attend the Corps, but only that which served to fasten their teeth.[50] Whether the *Opaline* stone in this Urne were burnt upon the finger of the dead, or cast into the fire by some affectionate friend, it will consist with either custome. But other incinerable substances were found so fresh, that they could feel no sindge from fire. These upon view were judged to be wood, but sinking in water and tried by the fire, we found them to be bone or Ivory. In their hardnesse and yellow colour they most resembled Box, which in old expressions found the Epithete[51] of Eternall, and perhaps in such conservatories* might have passed uncorrupted.

That Bay-leaves were found green in the Tomb of S. *Humbert*,[52] after an hundred and fifty years, was looked upon as miraculous. Remarkable it was unto old Spectators, that the Cypresse of the Temple of *Diana* lasted so many hundred years: The wood of the Ark and Olive Rod of *Aaron* were older at the Captivity. But the Cypresse of the Ark of *Noah* was the greatest vegetable Antiquity,* if *Josephus* were not deceived by some fragments of it in his dayes. To omit the Moore-logs, and Firre-trees found under-ground in many parts of *England*; the undated ruines of windes, flouds or earthquakes; and which in *Flanders* still shew from what quarter they fell, as generally lying in a North-East position.[53]

But though we found not these peeces to be Wood, according to first apprehension, yet we missed not altogether of some woody substance; For the bones were not so clearly pickt, but some coals were found amongst them; A way to make wood perpetuall, and a fit associat for metall, whereon was laid the foundation of the great *Ephesian* Temple, and which were made the lasting tests of old boundaries and Landmarks; Whilest we look on these, we admire not* Obser-

---

50. 12. *Tabul. l. xi de Jure Sacro. Neve aurum addito, ast quoi auro dentes vincti erunt, im cum illo sepelire & urere, se fraude esto.*

51. Plin. l. xvi. *Inter ξύλα ἀσαπῆ numerat Theophrastus.*

52. Surius.

53. Gorop. Becanus *in Niloscopio.*

vations of Coals found fresh, after four hundred years.[54] In a long deserted habitation,[55] even Egge-shels have been found fresh, not tending to corruption.

In the Monument of King *Childerick*, the Iron Reliques were found all rusty and crumbling into peeces. But our little Iron pins which fastened the Ivory works, held well together, and lost not their Magneticall quality, though wanting a tenacious moisture for the firmer union of parts; although it be hardly drawn into fusion, yet that metall soon submitteth unto rust and dissolution. In the brazen peeces we admired not the duration but the freedome from rust and ill savour, upon the hardest attrition; but now exposed unto the piercing Atomes of ayre, in the space of a few moneths, they begin to spot and betray their green entrals. We conceive not these Urnes to have descended thus naked as they appear, or to have entred their graves without the old habit of flowers.[56] The Urne of *Philopœmen* was so laden with flowers and ribbons, that it afforded no sight of it self. The rigid *Lycurgus* allowed Olive and Myrtle. The *Athenians* might fairly except against the practise of *Democritus* to be buried up in honey; as fearing to embezzle a great commodity of their Countrey, and the best of that kinde in *Europe*. But *Plato* seemed too frugally politick, who allowed no larger Monument than would contain four Heroick Verses, and designed* the most barren ground for sepulture: Though we cannot commend the goodnesse of that sepulchrall ground, which was set at no higher rate than the mean salary of *Judas*. Though the earth had confounded* the ashes of these Ossuaries, yet the bones were so smartly burnt, that some thin plates of brasse were found half melted among them; whereby we apprehend they were not of the meanest carcasses, perfunctorily fired as sometimes in military, and commonly in pestilence, burnings; or after the manner of abject corps, hudled forth and carelesly burnt, without the Esquiline Port at *Rome*; which was

54. Of Beringuccio *nella pyrotechnia*.
55. At Elmeham.
56. *ὑδρία* Plutarch.

an affront contrived upon *Tiberius*, while they but half burnt his body,[57] and in the Amphitheatre, according to the custome in notable Malefactors; whereas *Nero* seemed not so much to feare his death, as that his head should be cut off, and his body not burnt entire.

Some finding many fragments of sculs in these Urnes, suspected a mixture of bones; In none we searched was there cause of such conjecture, though sometimes they declined not that practise; The ashes of *Domitian*[58] were mingled with those of *Julia*, of *Achilles* with those of *Patroclus*: All Urnes contained not single Ashes; Without confused burnings they affectionately compounded their bones;* passionately endeavouring to continue their living Unions. And when distance of death denied such conjunctions, unsatisfied affections conceived some satisfaction to be neighbours in the grave, to lye Urne by Urne, and touch but in their names. And many were so curious to* continue their living relations, that they contrived large, and family Urnes, wherein the Ashes of their nearest friends and kindred might successively be received,[59] at least some parcels thereof, while their collaterall memorials lay in *minor* vessels about them.

Antiquity held too light thoughts from Objects of mortality, while some drew provocatives of mirth from Anatomies,[60] and Juglers shewed tricks with Skeletons. When Fidlers made not so pleasant mirth as Fencers, and men could sit with quiet stomacks while hanging was plaied before them.[61] Old considerations made few *memento's* by sculs and bones upon their monuments. In the Ægyptian Obelisks and Hieroglyphicall figures it is not easie to meet with

57. Sueton. in *vita Tib. Et in Amphitheatro semiustulanddum,* not. Casaub.

58. Sueton. in *vitâ Domitian.*

59. S. the most learned and worthy Mr M. Casaubon upon Antoninus.

60. *Sic erimus cuncti*, &c. *Ergo dum vivimus vivamus.*

61. *Ἀγχώνην παίζειν.* A barbarous pastime at Feasts, when men stood upon a rolling Globe, with their necks in a Rope fastned to a beame, and a knife in their hands, ready to cut it when the stone was rolled away, wherein if they failed they lost their lives to the laughter of their spectators. Athenæus.

bones. The sepulchrall Lamps speak nothing lesse than sepulture; and in their literall draughts prove often obscene and antick peeces: Where we finde *D.M.*[62] it is obvious to meet with* sacrificing *patera's,** and vessels of libation, upon old sepulchrall Monuments. In the Jewish *Hypogæum*[63]* and subterranean Cell at *Rome*, was little observable beside the variety of Lamps, and frequent draughts of the holy Candlestick. In authentick draughts of *Anthony* and *Jerome*, we meet with thigh-bones and deaths heads; but the cemiteriall Cels of ancient Christians and Martyrs, were filled with draughts of Scripture Stories; not declining the flourishes of Cypresse, Palmes, and Olive; and the mysticall Figures of Peacocks, Doves and Cocks. But iterately affecting the pourtraits of *Enoch*, *Lazarus*, *Jonas*, and the Vision of *Ezechiel*, as hopefull draughts, and hinting imagery of the Resurrection; which is the life of the grave, and sweetens our habitations in the Land of Moles and Pismires.*

Gentile Inscriptions precisely delivered the extent of mens lives, seldome the manner of their deaths, which history it self so often leaves obscure in the records of memorable persons. There is scarce any Philosopher but dies twice or thrice in *Laertius*; Nor almost any life without two or three deaths in *Plutarch*; which makes the tragicall ends of noble persons more favourably resented by compassionate Readers, who finde some relief in the Election of such differences.*

The certainty of death is attended with uncertainties, in time, manner, places. The variety of Monuments hath often obscured true graves: and *Cenotaphs** confounded Sepulchres. For beside their reall Tombs, many have founded honorary and empty Sepulchres. The variety of *Homers* Monuments made him of various Countreys. *Euripides*[64] had his Tomb in *Attica*, but his sepulture in *Macedonia*. And *Severus*[65] found his real Sepulchre in *Rome*, but his empty grave in *Gallia*.

62. *Diis manibus.*

63. Bosio.

64. Pausan. *in Atticis.*

65. Lamprid. in *vit. Alexand. Severi.*

He that lay in a golden Urne[66] eminently above the Earth, was not likely to finde the quiet of these bones. Many of these Urnes were broke by a vulgar discoverer in hope of inclosed treasure. The ashes of *Marcellus*[67] were lost above ground, upon the like account. Where profit hath prompted, no age hath wanted such miners. For which the most barbarous Expilators* found the most civill Rhetorick. Gold once out of the earth is no more due unto it; What was unreasonably committed to the ground is reasonably resumed from it: Let Monuments and rich Fabricks, not Riches adorn mens ashes. The commerce of the living is not to be transferred unto the dead: It is no injustice to take that which none complains to lose, and no man is wronged where no man is possessor.

What virtue yet sleeps in this *terra damnata* and aged cinders, were petty magick to experiment; These crumbling reliques and long-fired particles superannuate such expectations: Bones, hairs, nails, and teeth of the dead, were the treasures of old Sorcerers. In vain we revive such practices; Present superstition too visibly perpetuates the folly of our Fore-fathers, wherein unto old Observation[68] this Island was so compleat, that it might have instructed *Persia*.

Plato's historian of the other world* lies twelve dayes incorrupted, while his soul was viewing the large stations of the dead. How to keep the corps seven dayes from corruption by anointing and washing, without exenteration,* were an hazardable peece of art, in our choisest practise. How they made distinct separation of bones and ashes from fiery admixture, hath found no historicall solution. Though they seemed to make a distinct collection, and overlooked not *Pyrrhus* his toe.[69] Some provision they might make by fictile Vessels,* Coverings, Tiles, or flat stones, upon and about the body.

---

66. *Trajanus.* Dion.

67. Plut. in *vit. Marcelli.* The Commission of the Gothish King Theodoric for finding out sepulchrall treasure. *Cassiodor.* Var. I.4.

68. *Britannia hodie eam attonitè celebrat tantis ceremoniis, ut dedisse Persis videri possit.* Plin. 1. 30.

69. Which could not be burnt.

And in the same Field, not farre from these Urnes, many stones were found under ground, as also by carefull separation of extraneous matter, composing and raking up the burnt bones with forks, observable in that notable Lamp of *Galvanus*.[70] *Marlianus*,[71] who had the sight of the *Vas Ustrinum*, or vessell wherein they burnt the dead, found in the Esquiline Field at *Rome*, might have afforded clearer solution. But their insatisfaction herein begat that remarkable invention in the Funerall Pyres of some Princes, by incombustible sheets made with a texture of *Asbestos*, incremable* flax, or Salamanders wool, which preserved their bones and ashes incommixed.

How the bulk of a man should sink into so few pounds of bones and ashes, may seem strange unto any who considers not its constitution, and how slender a masse will remain upon an open and urging fire of the carnall composition. Even bones themselves reduced into ashes, do abate a notable proportion. And consisting much of a volatile salt, when that is fired out, make a light kind of cinders. Although their bulk be disproportionable to their weight, when the heavy principle of Salt is fired out, and the Earth almost only remaineth; Observable in sallow, which makes more Ashes than Oake; and discovers the common fraud of selling Ashes by measure, and not by ponderation.*

Some bones make best Skeletons,[72] some bodies quick and speediest ashes: Who would expect a quick flame from Hydropicall *Heraclitus?* The poysoned Souldier when his Belly brake, put out two pyres in *Plutarch*.[73] But in the plague of *Athens*,[74] one private pyre served two or three Intruders; and the *Saracens* burnt in large heaps,

70. To be seen in Licet. *de reconditis veterum lucernis*.

71. *Topographia Romæ ex Marliano. Erat & vas ustrinum appellatum quod in eo cadavera comburerentur. Cap. de Campo Esquilino.*

72. Old bones according to Lyserus. Those of young persons not tall nor fat according to Columbus.

73. In *vita Gracc.*

74. Thucydides.

by the King of *Castile*,[75] shewed how little Fuell sufficeth. Though the Funerall pyre of *Patroclus* took up an hundred foot,[76] a peece of an old boat burnt *Pompey*; And if the burthen of *Isaac* were sufficient for an holocaust, a man may carry his owne pyre.

From animals are drawn good burning lights, and good medicines against burning;[77] Though the seminall humour seems of a contrary nature to fire, yet the body compleated proves a combustible lump, wherein fire findes flame even from bones, and some fuell almost from all parts. Though the Metropolis[78] of humidity seems least disposed unto it, which might render the sculls of these Urnes lesse burned than other bones. But all flies or sinks before fire almost in all bodies: When the common ligament is dissolved, the attenuable parts ascend, the rest subside in coal, calx or ashes.

To burn the bones of the King of *Edom*[79] for Lyme, seems no irrationall ferity;* But to drink of the ashes of dead relations,[80] a passionate prodigality. He that hath the ashes of his friend, hath an everlasting treasure: where fire taketh leave, corruption slowly enters; In bones well burnt, fire makes a wall against it self; experimented in copels, and tests* of metals, which consist of such ingredients. What the Sun compoundeth, fire analyseth, not transmuteth. That devouring agent leaves almost allwayes a morsell for the Earth, whereof all things are but a colonie; and which, if time permits, the mother Element will have in their primitive masse again.

He that looks for Urnes and old sepulchrall reliques, must not seek them in the ruines of Temples; where no Religion anciently placed them. These were found in a Field, according to ancient custome, in noble or private buriall; the old practise of the *Canaanites*, the Family of *Abraham*, and the burying place of *Josua*, in the bor-

---

75. Laurent. Valla.

76. *Ἑκατόμπεδον ἔνθα και ἔνθα*. [Iliad. xxiii. 164]

77. *Sperm. ranarum*. Alb. Ovor.

78. The brain. Hippocrates.

79. Amos. 2. 1.*

80. As Artemisia of her husband Mausolus.

ders of his possessions; and also agreeable unto *Roman* practice to bury by high-wayes, whereby their Monuments were under eye: Memorials of themselves, and *memento's* of mortality unto living passengers; whom the Epitaphs of great ones were fain to beg to stay and look upon them. A language though sometimes used, not so proper in Church-Inscriptions.[81] The sensible Rhetorick of the dead, to exemplarity of good life, first admitted the bones of pious men and Martyrs within Church-wals; which in succeeding ages crept into promiscuous practise.* While *Constantine* was peculiarly favoured to be admitted unto the Church Porch; and the first thus buried in *England* was in the dayes of *Cuthred.*

Christians dispute how their bodies should lye in the grave. In urnall enterrment they clearly escaped this Controversie: Though we decline the Religious consideration, yet in cemiteriall and narrower burying places, to avoid confusion and crosse position, a certain posture were to be admitted; Which even Pagan civility observed.[82] The *Persians* lay North and South, The *Megarians* and *Phoenicians* placed their heads to the East: The *Athenians,* some think, towards the West, which Christians still retain. And *Beda* will have it to be the posture of our Saviour. That he was crucified with his face towards the West, we will not contend with tradition and probable account; but we applaud not the hand of the Painter, in exalting his Crosse so high above those on either side; since hereof we finde no authentick account in history, and even the crosses found by *Helena* pretend no such distinction from longitude or dimension.

To be gnaw'd out of our graves,* to have our sculs made drinking-bowls, and our bones turned into Pipes, to delight and sport our Enemies, are Tragicall abominations, escaped in burning Burials.

Urnall enterrments, and burnt Reliques lye not in fear of worms, or to be an heritage for Serpents; In carnall sepulture, corruptions seem peculiar unto parts, and some speak of snakes out of the spinall

---

81. Siste viator.

82. Kirckmannus *de funer.*

marrow. But while we suppose common wormes in graves, 'tis not easie to finde any there; few in Church-yards above a foot deep, fewer or none in Churches, though in fresh decayed bodies. Teeth, bones, and hair, give the most lasting defiance to corruption. In an Hydropicall body ten years buried in a Church-yard, we met with a fat concretion,* where the nitre of the Earth, and the salt and lixivious* liquor of the body, had coagulated large lumps of fat, into the consistence of the hardest castle-soap; whereof part remaineth with us. After a battle with the *Persians* the *Roman* Corps decayed in few dayes, while the *Persian* bodies remained dry and uncorrupted. Bodies in the same ground do not uniformly dissolve, nor bones equally moulder; whereof in the opprobrious disease* we expect no long duration. The body of the Marquesse of *Dorset* seemed sound and handsomely cereclothed, that after seventy eight years was found uncorrupted.[83] Common Tombs preserve not beyond powder: A firmer consistence and compage of parts might be expected from Arefaction,* deep buriall or charcoal. The greatest Antiquities of mortall bodies may remain in petrified bones, whereof, though we take not in the pillar of *Lots* wife, or Metamorphosis of *Ortelius,*[84] some may be older than Pyramids, in the petrified Reliques of the generall inundation.[85] When *Alexander* opened the Tomb of *Cyrus,* the remaining bones discovered his proportion, whereof urnall fragments afford but a bad conjecture, and have this disadvantage of grave enterrments, that they leave us ignorant of most personall discoveries. For since bones afford not only rectitude and stability, but figure unto the body; It is no impossible Physiognomy to conjecture at fleshy appendencies; and after what shape the muscles and carnous parts might hang in their full consistences. A full spread *Ca-*

83. Of Thomas, Marquesse of Dorset, whose body being buried 1530, was 1608 upon the cutting open of the Cerecloth found perfect and nothing corrupted, the flesh not hardened, but in colour, proportion, and softnesse like an ordinary corps newly to be interred. Burton's *descript. of Leicestershire.*

84. In his Map of Russia.

85. Wher in great numbers of men, oxen, and sheep were petrified.

riola[86] shews a well-shaped horse behinde, handsome formed sculls give some analogie of fleshy resemblance. A criticall view of bones makes a good distinction of sexes. Even colour is not beyond conjecture; since it is hard to be deceived in the distinction of *Negro's* sculls.[87] *Dantes*[88] Characters are to be found in sculls as well as faces. *Hercules* is not onely known by his foot. Other parts make out their comproportions, and inferences upon whole or parts. And since the dimensions of the head measure the whole body, and the figure thereof gives conjecture of the principall faculties; Physiognomy outlives our selves, and ends not in our graves.

Severe contemplators observing these lasting reliques, may think them good monuments of persons past, little advantage to future beings. And considering that power which subdueth all things unto it self, that can resume the scattered Atomes, or identifie* out of any thing, conceive it superfluous to expect a resurrection out of Reliques. But the soul subsisting, other matter clothed with due accidents may salve the individuality: Yet the Saints we observe arose from graves and monuments, about the holy City. Some think the ancient Patriarchs so earnestly desired to lay their bones in *Canaan,* as hoping to make a part of that Resurrection, and though thirty miles from Mount *Calvary,* at least to lie in that Region, which should produce the first-fruits of the dead. And if according to learned conjecture, the bodies of men shall rise where their greatest Reliques remain, many are not like to erre in the Topography of their Resurrection, though their bones or bodies be after translated

---

86. That part in the Skeleton of an Horse, which is made by the haunch-bones.

87. For their extraordinary thicknesse.

88. The Poet Dante in his view of Purgatory, found gluttons so meagre, and extenuated, that he conceited them to have been in the Siege of Jerusalem, and that it was easie to have discovered *Homo* or *Omo* in their faces: M being made by the two lines of their cheeks, arching over the Eyebrows to the nose, and their sunk eyes making O O which makes up *Omo.*

Parean l'occhiaie anella senza gemme:
Che nel viso degli huomini legge huomo,
Ben'havria quivi conosciuto l'emme.

by Angels into the field of *Ezechiels* vision, or as some will order it, into the Valley of Judgement, or *Jehosaphat*.[89]

CHAPTER IV

Christians have handsomely glossed the deformity of death, by careful consideration of the body, and civil rites which take off brutall terminations. And though they conceived all reparable by a resurrection, cast not off all care of enterrment. For since the ashes of Sacrifices burnt upon the Altar of God, were carefully carried out by the Priests, and deposed in a clean field; since they acknowledged their bodies to be the lodging of Christ, and temples of the holy Ghost, they devolved not all upon the sufficiency of soul existence;* and therefore with long services and full solemnities concluded their last Exequies, wherein to all distinctions the Greek devotion* seems most pathetically ceremonious.[90]

Christian invention hath chiefly driven at Rites, which speak hopes of another life, and hints of a Resurrection. And if the ancient Gentiles held not the immortality of their better part, and some subsistence after death; in severall rites, customes, actions and expressions, they contradicted their own opinions: wherein *Democritus* went high, even to the thought of a resurrection,[91] as scoffingly recorded by *Pliny*. What can be more expresse than the expression of *Phocyllides?*[92] Or who would expect from *Lucretius*[93] a sentence of *Ecclesiastes?* Before *Plato* could speak, the soul had wings in *Homer*, which fell not, but flew out of the body into the mansions of the dead; who also observed that handsome distinction of *Demas* and *Soma,** for the body conjoyned to the soul and body separated

---

89. *Tirin.* in Ezek.

90. *Rituale Græcorum opera J. Goar, in officio exequiarum.*

91. *Similis reviviscendi promissa a Democrito vanitas, qui non revixit ipse. Quæ malùm, ista dementia est; iterari vitam morte?* Plin. l. 7, c. 55.

92. *Καί τάχα δ' ἐκ γαίης ἐλπίζομεν ἐς φάος ἐλθεῖν λείψαν, ἀποιχωμένων, & deinceps.*

93. *Cedit enim retro de terra quod fuit ante. In terras*, &c. Lucret.

from it. *Lucian* spoke much truth in jest, when he said, that part of *Hercules* which proceeded from *Alchmena* perished, that from *Jupiter* remained immortall. Thus *Socrates*[94] was content that his friends should bury his body, so they would not think they buried *Socrates,* and regarding only his immortall part, was indifferent to be burnt or buried. From such Considerations *Diogenes* might contemn Sepulture. And being satisfied that the soul could not perish, grow carelesse of corporall enterrment. The *Stoicks* who thought the souls of wise men had their habitation about the *moon,* might make slight account of subterraneous deposition; whereas the *Pythagorians* and transcorporating Philosophers,* who were to be often buried, held great care of their enterrment. And the Platonicks rejected not a due care of the grave, though they put their ashes to unreasonable expectations, in their tedious term of return and long set revolution.

Men have lost their reason in nothing so much as their religion, wherein stones and clouts make Martyrs; and since the religion of one seems madnesse unto another, to afford an account or rationall of old Rites, requires no rigid Reader; That they kindled the pyre aversly, or turning their face from it, was an handsome Symbole of unwilling ministration; That they washed their bones with wine and milk, that the mother wrapt them in Linnen, and dryed them in her bosome, the first fostering part, and place of their nourishment; That they opened their eyes towards heaven, before they kindled the fire, as the place of their hopes or originall, were no improper Ceremonies. Their last valediction[95] thrice uttered by the attendants was also very solemn, and somewhat answered by Christians, who thought it too little, if they threw not the earth thrice upon the enterred body. That in strewing their Tombs the *Romans* affected the Rose, the Greeks *Amaranthus* and myrtle; that the Funerall pyre consisted of sweet fuell, Cypresse, Firre, Larix, Yewe, and Trees perpetually verdant, lay silent expressions of their surviving hopes:

94. Plato in *Phæd.*

95. *Vale, vale, vale, nos te ordine quo natura permittet sequemur.*

Wherein Christians which deck their Coffins with Bays have found a more elegant Embleme. For that tree seeming dead, will restore it self from the root, and its dry and exuccous* leaves resume their verdure again; which if we mistake not, we have also observed in furze. Whether the planting of yewe in Churchyards hold not its originall from ancient Funerall rites, or as an Embleme of Resurrection from its perpetual verdure, may also admit conjecture.

They made use of Musick to excite or quiet the affections of their friends, according to different harmonies. But the secret and symbolicall hint was the harmonical nature of the soul; which delivered from the body, went again to enjoy the primitive harmony of heaven, from whence it first descended; which according to its progresse traced by antiquity, came down by *Cancer,* and ascended by *Capricornus.*

They burnt not children before their teeth appeared, as apprehending their bodies too tender a morsell for fire, and that their gristly bones would scarce leave separable reliques after the pyrall combustion. That they kindled not fire in their houses for some dayes after, was a strict memoriall of the late afflicting fire. And mourning without hope, they had an happy fraud against excessive lamentation, by a common opinion that deep sorrows disturbed their ghosts.[96]

That they buried their dead on their backs, or in a supine position, seems agreeable unto profound sleep, and common posture of dying; contrary to the most naturall way of birth; nor like our pendulous posture, in the doubtfull state of the womb. *Diogenes* was singular, who preferred a prone situation in the grave, and some Christians[97] like neither, who decline the figure of rest, and make choice of an erect posture.

That they carried them out of the world with their feet forward, not inconsonant unto reason: As contrary unto the native posture of man, and his production first into it. And also agreeable unto their opinions, while they bid adieu unto the world, not to look again upon it; whereas *Mahometans** who think to return to a delightfull

96. *Tu manes ne læde meos.*

97. Russians, &c.

life again, are carried forth with their heads forward, and looking toward their houses.

They closed their eyes as parts which first die or first discover the sad effects of death. But their iterated clamations* to excitate their dying or dead friends, or revoke them unto life again, was a vanity of affection; as not presumably ignorant of the criticall tests of death, by apposition of feathers, glasses, and reflexion of figures, which dead eyes represent not; which however not strictly verifiable in fresh and warm *cadavers,* could hardly elude the test, in corps of four or five dayes.[98]

That they suck'd in the last breath of their expiring friends, was surely a practice of no medicall institution, but a loose opinion that the soul passed out that way, and a fondnesse of affection from some *Pythagoricall* foundation,[99] that the spirit of one body passed into another; which they wished might be their own.

That they powred oyle upon the pyre, was a tolerable practise, while the intention rested in facilitating the accension;* But to place good *Omens* in the quick and speedy burning, to sacrifice unto the windes for a dispatch in this office, was a low form of superstition.

The *Archimime* or *Jester* attending the Funerall train, and imitating the speeches, gesture, and manners of the deceased, was too light for such solemnities, contradicting their Funerall Orations, and dolefull rites of the grave.

That they buried a peece of money with them as a Fee of the *Elysian Ferry-man,* was a practise full of folly. But the ancient custome of placing coynes in considerable Urnes, and the present practise of burying medals in the Noble Foundations of *Europe,* are laudable wayes of historicall discoveries, in actions, persons, Chronologies; and posterity will applaud them.

We examine not the old Laws of Sepulture, exempting certain persons from buriall or burning. But hereby we apprehend that these were not the bones of persons Planet-struck or burnt with fire

98. At least by some difference from living eyes.

99. Francesco Perucci, *Pompe funebri.*

from Heaven: No Reliques of Traitors to their Countrey, Self-killers, or Sacrilegious Malefactors; Persons in old apprehension unworthy of the *earth*; condemned unto the *Tartarus* of Hell, and bottomlesse pit of *Pluto,* from whence there was no redemption.

Nor were only many customes questionable* in order to their Obsequies, but also sundry practises, fictions, and conceptions, discordant or obscure, of their state and future beings; whether unto eight or ten bodies of men to adde one of a woman, as being more inflammable, and unctuously constituted for the better pyrall combustion, were any rationall practise: Or whether the complaint of *Perianders* Wife be tolerable, that wanting her Funerall burning she suffered intolerable cold in Hell, according to the constitution of the infernall house of *Pluto,* wherein cold makes a great part of their tortures; it cannot passe without some question.

Why the Female Ghosts appear unto *Ulysses,* before the *Heroes* and masculine spirits? Why the *Psyche* or soul of *Tiresias* is of the masculine gender;[100] who being blinde on earth sees more than all the rest in hell; Why the Funerall Suppers consisted of Egges, Beans, Smallage, and Lettuce, since the dead are made to eat *Asphodels*[101] about the *Elyzian* medows? Why since there is no Sacrifice acceptable, nor any propitiation for the Covenant of the grave; men set up the Deity of *Morta,** and fruitlessly adored Divinities without ears? it cannot escape some doubt.

The dead seem all alive in the human *Hades* of *Homer,* yet cannot well speak, prophesie, or know the living, except they drink bloud, wherein is the life of man. And therefore the souls of *Penelope's* Paramours conducted by *Mercury* chirped like bats, and those which followed *Hercules* made a noise but like a flock of birds.

The departed spirits know things past and to come, yet are ignorant of things present. *Agamemnon* foretels what should happen unto *Ulysses,* yet ignorantly enquires what is become of his own Son. The Ghosts are afraid of swords in *Homer,* yet *Sybilla* tels *Æneas* in *Virgil,*

---

100. In Homer, *ψυχὴ Θηβαίον Τειρεσίαο σκῆπτρον ἔχων.*
101. In Lucian.

the thin habit of spirits was beyond the force of weapons. The spirits put off their malice with their bodies, and *Cæsar* and *Pompey* accord in Latine Hell, yet *Ajax* in *Homer* endures not a conference with *Ulysses*: And *Deiphobus* appears all mangled in *Virgils* Ghosts, yet we meet with perfect shadows among the wounded ghosts of *Homer*.

Since *Charon* in *Lucian* applauds his condition among the dead, whether it be handsomely said of *Achilles,* that living contemner of death, that he had rather be a Plowmans servant than Emperour of the dead? How *Hercules* his soul is in hell, and yet in heaven, and *Julius* his soul in a Starre, yet seen by *Æneas* in hell, except the Ghosts were but Images and shadows of the soul, received in higher mansions, according to the ancient division of body, soul, and image or *simulachrum* of them both. The particulars of future beings must needs be dark unto ancient Theories, which Christian Philosophy yet determines but in a Cloud of opinions. A Dialogue between two Infants in the womb concerning the state of this world, might handsomely illustrate our ignorance of the next, whereof methinks we yet discourse in *Platoes* denne, and are but *Embryon* Philosophers.

Pythagoras escapes in the fabulous hell of *Dante,*[102] among that swarm of Philosophers, wherein whilest we meet with *Plato* and *Socrates, Cato* is to be found in no lower place than Purgatory. Among all the set, *Epicurus* is most considerable, whom men make honest without an *Elyzium,* who contemned life without encouragement of immortality, and making nothing after death, yet made nothing of the King of terrours.

Were the happinesse of the next world as closely apprehended as the felicities of this, it were a martyrdome to live; and unto such as consider none hereafter, it must be more than death to dye, which makes us amazed at those audacities, that durst be nothing, and return into their *Chaos* again. Certainly such spirits as could contemn death, when they expected no better being after, would have scorned to live had they known any. And therefore we applaud not the judgment of *Machiavel,* that Christianity makes men cowards,

102. *Del inferno,* cant. 4.

or that with the confidence of but half dying, the despised virtues of patience and humility have abased the spirits of men, which Pagan principles exalted, but rather regulated the wildenesse of audacities, in the attempts, grounds, and eternall sequels of death; wherein men of the boldest spirits are often prodigiously temerarious. Nor can we extenuate the valour of ancient Martyrs,* who contemned death in the uncomfortable scene of their lives, and in their decrepit Martyrdomes did probably lose not many moneths of their dayes, or parted with life when it was scarce worth the living. For (beside that long time past holds no consideration unto a slender time to come) they had no small disadvantage from the constitution of old age, which naturally makes men fearfull; complexionally superannuated from the bold and couragious thoughts of youth and fervent years. But the contempt of death from corporall animosity* promoteth not our felicity. They may sit in the *Orchestra*, and noblest Seats of Heaven, who have held up shaking hands in the fire, and humanly contended for glory.

Mean while *Epicurus* lyes deep in *Dante's* hell, wherein we meet with Tombs enclosing souls which denied their immortalities. But whether the virtuous heathen, who lived better than he spake, or erring in the principles of himself, yet lived above Philosophers of more specious Maximes, lye so deep as he is placed; at least so low as not to rise against Christians, who beleeving or knowing that truth, have lastingly denied it in their practise and conversation, were a quæry too sad to insist on.

But all or most apprehensions rested in Opinions of some future being, which ignorantly or coldly beleeved, begat those perverted conceptions, Ceremonies, Sayings, which Christians pity or laugh at. Happy are they, which live not in that disadvantage of time, when men could say little for futurity, but from reason. Whereby the noblest mindes fell often upon doubtfull deaths, and melancholly Dissolutions; With these hopes *Socrates* warmed his doubtfull spirits against that cold potion, and *Cato* before he durst give the fatall stroak spent part of the night in reading the immortality of *Plato*,* thereby confirming his wavering hand unto the animosity of that attempt.

It is the heaviest stone that melancholy can throw at a man, to tell him he is at the end of his nature; or that there is no further state to come, unto which this seemes progressionall, and otherwise made in vaine; Without this accomplishment the naturall expectation and desire of such a state, were but a fallacy in nature; unsatisfied Considerators would quarrell the justice of their constitutions, and rest content that *Adam* had fallen lower, whereby by knowing no other Originall, and deeper ignorance of themselves, they might have enjoyed the happinesse of inferiour Creatures; who in tranquility possesse their Constitutions, as having not the apprehension to deplore their own natures. And being framed below the circumference of these hopes, or cognition of better being, the wisedom of God hath necessitated their Contentment: But the superiour ingredient and obscured part of our selves, whereto all present felicities afford no resting contentment, will be able at last to tell us we are more than our present selves; and evacuate such hopes in the fruition of their own accomplishments.

CHAPTER V

Now since these dead bones have already out-lasted the living ones of *Methuselah*, and in a yard under ground, and thin walls of clay, out-worn all the strong and specious buildings above it; and quietly rested under the drums and tramplings of three conquests; What Prince can promise such diuturnity* unto his Reliques, or might not gladly say,

Sic ego componi versus in ossa velim.[103*]

Time which antiquates Antiquities, and hath an art to make dust of all things, hath yet spared these *minor* Monuments. In vain we hope to be known by open and visible conservatories, when to be un-

---

103. Tibullus.

known was the means of their continuation and obscurity their protection: If they dyed by violent hands, and were thrust into their Urnes, these bones become considerable, and some old Philosophers would honour them,[104] whose souls they conceived most pure, which were thus snatched from their bodies; and to retain a stronger propension* unto them: whereas they weariedly left a languishing corps, and with faint desires of re-union. If they fell by long and aged decay, yet wrapt up in the bundle of time, they fall into indistinction, and make but one blot with Infants. If we begin to die when we live, and long life be but a prolongation of death, our life is a sad composition; We live with death, and die not in a moment. How many pulses made up the life of *Methuselah*, were work for *Archimedes*: Common Counters summe up the life of *Moses* his man.[105] Our dayes become considerable like petty sums by minute accumulations; where numerous fractions make up but small round numbers; and our dayes of a span long make not one little finger.[106]

If the nearnesse of our last necessity brought a nearer conformity unto it, there were a happinesse in hoary hairs, and no calamity in half senses. But the long habit of living indisposeth us for dying; When Avarice makes us the sport of death; When even *David* grew politickly cruell; and *Solomon* could hardly be said to be the wisest of men. But many are too early old, and before the date of age. Adversity stretcheth our dayes, misery makes *Alcmenas* nights,[107] and time hath no wings unto it. But the most tedious being is that which can unwish it self, content to be nothing, or never to have been, which was beyond the *male*-content of *Job*, who cursed not the day of his life, but his Nativity: Content to have so farre been, as to have a Title to future being; Although he had lived here but in an hidden state of life, and as it were an abortion.

---

104. *Oracula Chaldaica cum scholiis Pselli & Plethonis. βίῃ λιπόντων σῶμα ψυχαὶ καθαρώταται. Vi Corpus reliquentium animæ Urisimsæ.*
105. In the Psalme of *Moses*.
106. According to the ancient Arithmetick of the hand wherein the little finger of the right hand contracted, signified an hundred. Pierius in *Hieroglyph*.
107. One night as long as three.

What Song the *Syrens* sang, or what name *Achilles* assumed when he hid himself among women, though puzling Questions[108] are not beyond all conjecture. What time the persons of these Ossuaries entred the famous Nations of the dead,[109] and slept with Princes and Counsellours,[110] might admit a wide solution.* But who were the proprietaries of these bones, or what bodies these ashes made up, were a question above Antiquarism. Not to be resolved by man, nor easily perhaps by spirits, except we consult the Provinciall Guardians, or tutellary Observators.* Had they made as good provision for their names, as they have done for their Reliques, they had not so grosly erred in the art of perpetuation. But to subsist in bones, and be but Pyramidally extant, is a fallacy in duration. Vain ashes, which in the oblivion of names, persons, times, and sexes, have found unto themselves a fruitlesse continuation, and only arise unto late posterity, as Emblemes of mortall vanities; Antidotes against pride, vainglory, and madding vices. Pagan vain-glories which thought the world might last for ever, had encouragement for ambition, and finding no *Atropos* unto the immortality of their Names, were never dampt with the necessity of oblivion. Even old ambitions had the advantage of ours, in the attempts of their vainglories, who acting early, and before the probable Meridian of time,* have by this time found great accomplishment of their designes, whereby the ancient *Heroes* have already outlasted their Monuments, and Mechanicall preservations. But in this latter Scene of time we cannot expect such Mummies unto our memories,* when ambition may fear the Prophecy of *Elias*,[111] and *Charles* the fifth can never hope to live within two *Methusela's* of *Hector*.[112]

And therefore restlesse inquietude for the diuturnity of our memories unto present considerations seems a vanity almost out of date, and superanuated peece of folly. We cannot hope to live so long

108. The puzling questions of Tiberius unto grammarians. *Marcel. Donatus* in Suet.
109. *κλυτα ἔθνεα*. Hom.
110. Job.
111. That the world may last but six thousand years.
112. Hector's fame lasting above two lives of Methuselah, before that famous Prince was extant.

in our names as some have done in their persons, one face of *Janus* holds no proportion unto the other. 'Tis too late to be ambitious. The great mutations of the world are acted, our time may be too short for our designes. To extend our memories by Monuments, whose death we dayly pray for, and whose duration we cannot hope, without injury to our expectations in the advent of the last day, were a contradiction to our beliefs. We whose generations are ordained in this setting part of time, are providentially taken off from such imaginations. And being necessitated to eye the remaining particle of futurity, are naturally constituted unto thoughts of the next world, and cannot excusably decline the consideration of that duration, which maketh Pyramids pillars of snow, and all that's past a moment.

Circles and right lines limit and close all bodies, and the mortall right-lined circle[113] must conclude and shut up all. There is no antidote against the *Opium* of time, which temporally considereth all things; Our Fathers finde their graves in our short memories, and sadly tell us how we may be buried in our Survivors. Grave-stones tell truth scarce fourty years:[114] Generations passe while some trees stand, and old Families last not three Oaks. To be read by bare Inscriptions like many in *Gruter*,[115] to hope for Eternity by Ænigmaticall Epithetes, or first letters of our names, to be studied by Antiquaries, who we were, and have new Names given us like many of the Mummies,[116] are cold consolations unto the Students of perpetuity, even by everlasting Languages.

To be content that times to come should only know there was such a man, not caring whether they knew more of him, was a frigid ambition in *Cardan*:[117] disparaging his horoscopal inclination and judgement of himself. Who cares to subsist like *Hippocrates* Patients, or *Achilles* horses in *Homer*, under naked nominations, with-

---

113. *Θ* The Character of death.

114. Old ones being taken up, and other bodies laid under them.

115. *Gruteri Inscriptiones Antiquæ.*

116. Which men show in several Countries, giving them what names they please; and unto some the names of the old Ægyptian Kings out of Herodotus.

117. *Cuperem notum esse quod sim, non opto ut sciatur qualis sim.* Card. in *vita propria.*

out deserts and noble acts, which are the balsame of our memories, the *Entelechia** and soul of our subsistences? To be namelesse in worthy deeds exceeds an infamous history. The *Canaanitish* woman* lives more happily without a name, than *Herodias* with one. And who had not rather have been the good theef, than *Pilate?*

But the iniquity of oblivion blindely scattereth her poppy, and deals with the memory of men without distinction to merit of perpetuity. Who can but pity the founder of the Pyramids? *Herostratus* lives that burnt the Temple of *Diana*, he is almost lost that built it; Time hath spared the Epitaph of *Adrians* horse, confounded that of himself. In vain we compute our felicities by the advantage of our good names, since bad have equall durations; and *Thersites* is like to live as long as *Agamemnon*. Who knows whether the best of men be known? or whether there be not more remarkable persons forgot, than any that stand remembred in the known account of time? Without the favour of the everlasting Register* the first man had been as unknown as the last, and *Methuselahs* long life had been his only Chronicle.

Oblivion is not to be hired: The greater part must be content to be as though they had not been, to be found in the Register of God, not in the record of man. Twenty seven Names make up the first story,[118] and the recorded names ever since contain not one living Century. The number of the dead long exceedeth all that shall live. The night of time far surpasseth the day, and who knows when was the Æquinox? Every houre addes unto that current Arithmetique, which scarce stands one moment. And since death must be the *Lucina* of life, and even Pagans could doubt[119] whether thus to live, were to dye. Since our longest Sunne sets at right descensions, and makes but winter arches, and therefore it cannot be long before we lie down in darknesse, and have our light in ashes.[120] Since the

118. Before the flood.

119. Euripides [Polyidos].

120. According to the custome of the Jewes, who place a lighted wax-candle in a pot of ashes by the corps. Leo [of Modena].

brother of death daily haunts us with dying *memento's*, and time that grows old it self, bids us hope no long duration: Diuturnity is a dream and folly of expectation.

Darknesse and light divide the course of time, and oblivion shares with memory a great part even of our living beings; we slightly remember our felicities, and the smartest stroaks of affliction leave but short smart upon us. Sense endureth no extremities, and sorrows destroy us or themselves. To weep into stones are fables. Afflictions induce callosities,* miseries are slippery, or fall like snow upon us, which notwithstanding is no unhappy stupidity. To be ignorant of evils to come, and forgetfull of evils past, is a mercifull provision in nature, whereby we digest the mixture of our few and evil dayes, and our delivered senses not relapsing into cutting remembrances, our sorrows are not kept raw by the edge of repetitions. A great part of Antiquity contented their hopes of subsistency with a transmigration of their souls. A good way to continue their memories, while having the advantage of plurall successions, they could not but act something remarkable in such variety of beings, and enjoying the fame of their passed selves, make accumulation of glory unto their last durations. Others rather than be lost in the uncomfortable night of nothing, were content to recede into the common being, and make one particle of the publick soul of all things, which was no more than to return into their unknown and divine Originall again. Ægyptian ingenuity was more unsatisfied, continuing their bodies in sweet consistences, to attend the return of their souls. But all was vanity, feeding the winde,[121] and folly. The Ægyptian Mummies, which *Cambyses* or time hath spared, avarice now consumeth. Mummie is become Merchandise,* *Miszraim* cures wounds, and *Pharaoh* is sold for balsoms.

In vain do individuals hope for Immortality, or any patent from oblivion, in preservations below the Moon: Men have been deceived even in their flatteries above the Sun, and studied conceits to perpetuate their names in heaven. The various Cosmography of that

121. *Omnia vanitas & pastio, venti, νομὴ ἀνέμον βόσκησις ut olim Aquila & Symmachus* V. Drus. *Eccles.*

part hath already varied the names of contrived constellations; *Nimrod* is lost in *Orion,* and *Osyris* in the Dogge-starre. While we look for incorruption in the heavens, we finde they are but like the Earth; Durable in their main bodies, alterable in their parts: whereof beside Comets and new Stars, perspectives* begin to tell tales. And the spots that wander about the Sun, with *Phaetons* favour, would make clear conviction.

There is nothing strictly immortall, but immortality. Whatever hath no beginning may be confident of no end (all others have a dependent being, and within the reach of destruction) which is the peculiar of that necessary essence that cannot destroy it self; And the highest strain of omnipotency to be so powerfully constituted, as not to suffer even from the power of it self. But the sufficiency of Christian Immortality frustrates all earthly glory, and the quality of either state after death makes a folly of posthumous memory. God who only can destroy our souls, and hath assured our resurrection, either of our bodies or names hath directly promised no duration. Wherein there is so much of chance that the boldest Expectants have found unhappy frustration; and to hold long subsistence, seems but a scape in oblivion. But man is a Noble Animal, splendid in ashes, and pompous in the grave, solemnizing Nativities and Deaths with equall lustre, nor omitting Ceremonies of bravery, in the infamy of his nature.

Life is a pure flame, and we live by an invisible Sun within us. A small fire sufficeth for life, great flames seemed too little after death, while men vainly affected precious pyres, and to burn like *Sardanapalus*; but the wisedom of funerall Laws found the folly of prodigall blazes, and reduced undoing fires unto the rule of sober obsequies, wherein few could be so mean as not to provide wood, pitch, a mourner, and an Urne.[122]

Five languages secured not the Epitaph of *Gordianus*;[123] The man

---

122. According to the Epitaph of Rufus and Beronica in Gruterus.... *Nec ex Eorum bonis plus inventum est, quam Quod sufficeret ad emendam pyram, Et picem quibus corpora cremarentur, Et præfica conducta & olla empta.*

123. In Greek, Latine, Hebrew, Ægyptian, Arabick, defaced by Licinius the Emperour.

of God lives longer without a Tomb than any by one, invisibly interred by Angels, and adjudged to obscurity though not without some marks directing human discovery. *Enoch* and *Elias* without either tomb or buriall, in an anomalous state of being, are the great Examples of perpetuity in their long and living memory, in strict account being still on this side death, and having a late part yet to act upon this stage of earth. If in the decretory term of the world* we shall not all dye but be changed, according to received translation, the last day will make but few graves; at least quick Resurrections will anticipate lasting Sepultures; Some Graves will be opened before they be quite closed, and *Lazarus* be no wonder. When many that feared to dye shall groane that they can dye but once, the dismall state is the second and living death; when life puts despair on the damned; when men shall wish the coverings of Mountaines, not of Monuments, and annihilation shall be courted.

While some have studied Monuments, others have studiously declined them: and some have been so vainly boisterous, that they durst not acknowledge their Graves; wherein[124] *Alaricus* seems most subtle, who had a River turned to hide his bones at the bottome. Even *Sylla* that thought himself safe in his Urne, could not prevent revenging tongues, and stones thrown at his Monument. Happy are they whom privacy makes innocent, who deal so with men in this world, that they are not afraid to meet them in the next, who when they dye, make no commotion among the dead, and are not toucht with that poeticall taunt of *Isaiah*.[125]

Pyramids, Arches, Obelisks, were but the irregularities of vainglory, and wilde enormities of ancient magnanimity. But the most magnanimous resolution rests in the Christian Religion, which trampleth upon pride, and sits on the neck of ambition, humbly pursuing that infallible perpetuity, unto which all others must diminish their diameters, and be poorly seen in Angles of contingency.[126]

124. Jornandes *de rebus Geticis.*

125. Isa. 14.

126. *Angulus contingentiæ,* the least of Angles.

Pious spirits who passed their dayes in raptures of futurity, made little more of this world than the world that was before it, while they lay obscure in the Chaos of pre-ordination, and night of their fore-beings. And if any have been so happy as truly to understand Christian annihilation, extasis, exolution,* liquefaction, transformation, the kisse of the Spouse, gustation of God, and ingression into the divine shadow, they have already had an handsome anticipation of heaven; the glory of the world is surely over, and the earth in ashes unto them.

To subsist in lasting Monuments, to live in their productions, to exist in their names, and prædicament of *Chymera's*,* was large satisfaction unto old expectations, and made one part of their *Elyziums*. But all this is nothing in the Metaphysicks of true belief. To live indeed is to be again our selves, which being not only an hope but an evidence in noble beleevers, 'Tis all one to lye in St *Innocents*[127] Church-yard, as in the Sands of *Ægypt*: Ready to be any thing, in the extasie of being ever, and as content with six foot as the Moles of *Adrianus*.[128]

Lucan

—Tabesne cadavera solvat
*An rogus haud refert.**—

127. In Paris where bodies soon consume.

128. A stately *Mausoleum* or sepulchral pyle built by Adrianus in Rome, where now standeth the Castle of St. Angelo.

A PASSAGE FROM A MANUSCRIPT

The following lines, closely related to the paragraph that begins at the top of page 136, are included in both Simon Wilkin's and Geoffrey Keynes's editions of Browne's works.

Large are the treasures of oblivion, and heapes of things in a state next to nothing almost numberlesse; much more is buried in silence than is recorded, and the largest volumes are butt epitomes of what hath been. The account of time beganne with night, and darknesse still attendeth it. Some things never come to light; many have been delivered; butt more hath been swallowed in obscurity & the caverns of oblivion. How much is as it were *in vacuo*, and will never be cleered up, of those long living times when men could scarce remember themselves young; and men seeme to us not ancient butt antiquities; when they subsisted longer in their lives then wee can now hope to do in our memories; when men feared apoplexies & palsies after 7 or 8 hundred years; when living was so lasting that homicide might admitt of distinctive qualifications from the age of the person, & it might seeme a lesser offense to kill a man at 8 hundred then at fortie, and when life was so well worth the living that few or none would kill themselves.

GLOSSARY OF NAMES AND PLACES

Absalom: Son of King David, rebelled against his father and was killed in battle (2 Samuel: 13–18).

Adrian (Hadrian): (76–138 CE) Roman emperor.

Aelian: (3rd century CE) Roman author whose works on animals are known for their incredible lore.

Agricola: (40–93 CE) Roman general responsible for much of the conquest of Britain.

Ajax (Telamon): One of the most formidable Greek warriors in Homer's *Iliad*.

Ajax Oileus: One of the Greek warriors in Homer's *Iliad*.

Alaricus (Alaric): (c. 370–410 CE) Ruler of the Visigoths, conqueror of Rome.

Alchmena (Alcmene): Mother of Hercules in Greek mythology.

Anaxagoras: (c. 500–428 BCE) Greek philosopher, known for his theory of multiple worlds.

Ansgarius: (801–65 CE) Archbishop of Hamburg-Bremen.

Anthony: (c. 251–356 CE) Christian saint and hermit.

Antoninus: (86–161 CE) Roman emperor.

Archemorus: In Greek mythology, a child of the King of Nemea whose death was honored with a vast funeral pyre.

Archimedes: (c. 287–212 BCE) Greek mathematician and engineer.

Arius: (250–336 CE) Alexandrian Christian theologian, founder of the heresy denying Christ's divinity and consubstantiality with God the Father.

Asa: (c. 910–869 BCE) Third King of Judah.

Atropos: One of the three Fates of Greek mythology, Atropos cuts the thread of life.

Baldwin: (c. 1058–1118) First king of the Crusader state of Jerusalem.

Balearians: Ancient inhabitants of the Balearic Islands, an archipelago off the eastern coast of the Iberian peninsula.

Beda (Bede): (c. 673–735 CE) English monk, scholar, and historian.

Bellinus: Legendary king of the Britons.

Boadicea: (c. 1st century CE) Queen of the Iceni, led a revolt against the Roman forces in Britain.

Bosio, Antonio: (c. 1575–1629) Italian scholar and archaeologist.

Britannicus: (41–55 CE) Son of the emperor Claudius.

Cambyses II: (d. 530 BCE) Persian emperor.

Canutus (Canute II): (d. 1035 CE) King of England and Denmark.

Caracalla: (188–217 CE) Roman emperor.

Cardano, Girolamo: (1501–1576) Italian physician, mathematician, and astrologer.

Casalius, Joannes Baptista: (17th century) Italian antiquary and historian.

Cato (the Younger): (95–46 BCE) Roman orator and statesman, guardian of Purgatory in Dante's *Divine Comedy*.

Chaldeans: Ancient inhabitants of the Babylonian Empire, averse to the practice of cremation.

Charles V: (1500–1558) Holy Roman emperor.

Charon: Ferryman of Hades in Greek mythology.

Childeric I: (c. 440–481 CE) King of the Salian Franks.

Claudius: (10 BCE–54 CE) Roman emperor.

Codrus: (11th century BCE) Legendary king of Athens, he sacrificed himself to save the city from the invading Dorians.

Commodus: (161–192 CE) Roman emperor.

Constans: (323–350 CE) Roman emperor.

Constantine: (272–337 CE) First Roman emperor to convert to Christianity.

Cornelius Sylla: (c. 138–78 BCE) Roman general and statesman, his body was cremated in Rome.

Curtius, Marcus: (4th century BCE) Roman knight, according to legend sacrificed himself to save Rome.

Cuthred: (d. 754 CE) King of the West Saxons.

Cyrus: (c. 600–530 BCE) Founder of the Persian Empire.

Deiphobus: Trojan prince, brother of Hector and Paris.

Democritus: (c. 460–370 BCE) Greek philosopher, known as the "laughing philosopher" for his tendency to mock human folly.

Dioclesian (Diocletian): (244–311 CE) Roman emperor.

Diogenes: (c. 412–323 BCE) Greek philosopher and famous Cynic.

Domitian: (51–96 CE) Roman emperor.

Eliah (Elijah): Hebrew prophet whose translation to heaven was considered a type or anticipation of the resurrection.

Enoch: Biblical patriarch who was taken to God while still living, often seen as a type or prefiguration of Christ.

Epicurus: (c. 341–270 BCE) Greek materialist philosopher.

Farnese: (1520–1589) Italian cardinal, diplomat, and famous collector and patron of the arts.

Frotho: Legendary Danish king.

Gaguinus (Alexander Guagnini): (c. 1538–1614) Italian historian.

Galen: (2nd century CE) Roman physician (of Greek origins), most renowned medical writer of antiquity.

Gallienus: (218–268 CE) Roman emperor whose reign was notable for the great number of pretenders to the throne (the "thirty tyrants").

Geta: (189–211 CE) Roman emperor.

Gordianus III: (225–244 CE) Roman emperor.

Gruter, Jan: (1560–1627) Dutch scholar, known for his collection of ancient inscriptions.

Harald: (c. 8th century) Legendary Scandinavian king.

Helena: (255–330 CE) Mother of the Roman emperor Constantine.

Heliogabalus: (203–222) Roman emperor.

Heraclitus: (c. 535–475 BCE) Greek philosopher who proposed fire as the fundamental element of the universe, known as "the weeping philosopher."

Hermes Trismegistus: Legendary author of a large body of hermetic writings.

Herodias: Wife of Herod Antipas in the Gospel of Mark.

Herodotus: (484–425 BCE) Greek historian.

Herostratus: Greek who burned the famous Temple of Diana in 356 BCE.

Hippocrates: (c. 460–370 BCE) Famous Greek physician.

Humbert (Saint Hubert): (d. 727) Bishop of Liège.

Husse, John (Jan Hus): (1369–1415) Early Bohemain Reformer, burned at the stake for heresy.

Iceni: Ancient inhabitants of modern-day Norfolk.

Jair: Eighth Judge or ruler of Israel, according to Browne a contemporary of Archemorus.

Janus: Two-faced Roman god.

Jehoram: (c. 9th century BCE) King of Judah.
Jerome: (c. 342–420 CE) Church father and translator of the Bible.
Jonah: Hebrew prophet, lived for three days in the belly of a whale. His story was read by Christians as a prefiguration of Christ's death and resurrection.
Josephus, Flavius: (c. 37–100 CE) Jewish historian.
Julian: (331–363 CE) Roman emperor.
Justine (Justin): (c. 3rd century CE) Roman historian.
Laertius, Diogenes: (2nd or 3rd century CE) Greek author of the biographies of several eminent philosophers.
Lucan: (39–65 CE) Roman poet.
Lucian: (c. 125–180 CE) Greek satirist and rhetorician.
Lucina: Roman goddess of childbirth.
Lucretius: (c. 99–55 BCE) Roman poet and materialist philosopher.
Ludovicus Pius: (778–840 CE) King of the Franks.
Lycurgus: Legendary Spartan lawgiver.
Macrobius: (fl. 395–423 CE) Roman grammarian and philosopher.
Manlius: (d. 384 BCE) Roman consul.
Marcellus: (c. 268–208 BCE) Roman general and statesman.
Marcus Aurelius: (121–180) Roman emperor.
Marius: (c. 157–86 BCE) Roman general and statesman.
Marlianus, Joannes Bartholomaeus: (c. 1490–1560) Italian antiquary and topographer.
Matilda: (1102–1167) Holy Roman empress, daughter of Henry I of England and mother of Henry II.
Mausolus: (d. 353 BCE) King of Caria, buried in an extravagant tomb.
Megasthenes: (c. 350–290 BCE) Greek historian and ethnographer.
Menecus (Menoeceus): In Greek mythology, Menecus dies in the battle for Thebes and was (according to Browne) cremated.
Methuselah: Biblical patriarch, lived to the age of 969.
Minucius: (c. 2nd or 3rd century CE) Roman apologist for Christianity.
Morpheus: In Greek mythology, the god of sleep and dreams.
Nero: (37–68 CE) Roman emperor.
Nimrod: Great-grandson of Noah and legendary founder of the Assyrian Empire.
Numa: Legendary second king of Rome.
Olaus Magnus: (c. 1490–1557) Swedish historian.

Origen: (185–254 CE) Early Alexandrian Christian theologian.
Ortelius, Abraham: (1527–1598) Flemish cartographer.
Ostorius: (d. 52 CE) Roman general and statesman.
Osyris (Osiris): Egyptian god of the underworld.
Paracelsus: (1493–1541) Swiss alchemist, physician, and astrologer.
Patroclus: Companion of Achilles in Homer's *Iliad*, burned on a great funeral pyre.
Paulinus: (1st century CE) Roman general who quelled the rebellion of Boadicea.
Penthisilea: Queen of the Amazonians in Homer's *Iliad*.
Periander: Ruler of Corinth, one of the Seven Sages of ancient world.
Phaeton: Son of Helios in Greek mythology, struck down by Zeus while attempting to drive the chariot of the sun.
Philo: (20 BCE–50 CE) Jewish biblical interpreter.
Philopoemen: (253–183 BCE) Greek general and statesman.
Phocylides: (6th century BCE) Greek poet and philosopher.
Pineda, Juan de: (16th century) Spanish theologian.
Pliny the Elder: (c. 23–79 CE) Roman natural philosopher.
Plutarch: (c. 46–120 CE) Greek historian and biographer.
Polydorus (Polydore Vergil): (c. 1470–1555) Italian historian of England.
Pompey: (106–48 BC) Roman military and political leader, defeated by Caesar at the battle of Pharsalus.
Pomponius: (1st century CE) Roman geographer.
Poppaea: (30–65 BCE) Wife of Nero, elaborately embalmed and entombed after her death.
Porphyry: (c. 234–305 CE) Neoplatonist philosopher.
Prasutagus: (c. 1st century CE) King of the Iceni and husband of Boadicea.
Propertius: (d. 15 BCE) Roman elegiac poet.
Ptolemy II: (d. 246 BCE) King of Egypt, supposed to have commissioned the Septuagint for his library in Alexandria.
Pythagoras: (c. 570–495 BCE) Greek philosopher and mathematician.
Regio-Montanus (Johannes Müller von Königsberg): (15th century) Mathematician, astronomer, and inventor.
Remus: Along with his twin brother, Romulus, legendary founder of Rome.
Rhadamanth: According to Greek mythology, one of the judges of the dead.
Ringo: (c. 8th century CE) Legendary Scandinavian king.

Rollo: (c. 846–931) Norse warrior and first Duke of Normandy.

Sardanapalus: (7th century BCE) Legendary king of Assyria.

Saxo (Grammaticus): (c. 1150–1220) Danish historian.

Scevola, Gaius Mucius: (6th century BCE) Legendary Roman warrior, burned his own hand to illustrate the bravery of Romans to their Etruscan enemies.

Sedechias (Zedekiah): (c. 618–587 BCE) Last king of Judah before its destruction by Babylon.

Severus: (145–211 CE) Roman emperor.

Sidonius Apollinaris: (c. 430–489 CE) Gallo-Roman poet, writer, and bishop.

Starkatterus (Starkaor): Warrior of Scandinavian legend.

Strabo: (c. 64 BCE–24 CE) Greek historian and geographer.

Suarez, Francisco: (1548–1617) Spanish theologian.

Sueno (Sven or Sweyn): (d. 1014) King of Denmark and conqueror of England.

Sybilla (Sibyl): Prophetess or soothsayer.

Tacitus: (c. 56–177 CE) Roman historian.

Tarquinus Priscus: (616–579 BCE) Fifth king of Rome.

Tertullian: (c. 160–220 CE) Early Christian theologian.

Thales: (c. 624–546 BCE) Greek philosopher who proposed water as the fundamental element of the universe.

Thersites: The misshapen and vulgar soldier in Homer's *Iliad*.

Tiberius: (42 BCE–37 CE) Roman emperor.

Tiresias: Blind prophet of Thebes.

Titus: (39–81 CE) Roman emperor.

Trajan: (53–117 CE) Roman emperor.

Tymon (Timon): (5th century BCE) Famous Athenian misanthrope.

Ulfketel (Ulfkell): (died 1016) Leader of the East Anglian forces opposing Sueno.

Unguinus: Legendary Swedish king.

Valens: (328–378 CE) Emperor of the Eastern Roman Empire.

Varro: (116–27 BCE) Roman scholar and writer.

Venta: Market town of the Iceni in modern-day Norfolk.

Vespasian: (9–79 CE) Roman emperor.

Walsingham: An English village in the north of Norfolk.

William (the Conqueror): (1027–1087) Norman duke, later king of England.
Wormius: (1588–1655) Danish physician and antiquary.
Zeno: (335–263 BCE) philosopher, founder of Stoicism.
Zoroaster: Ancient Persian prophet, founder of Zoroastrianism.

NOTES

RELIGIO MEDICI

3 *that excellent invention:* The printing press.

most imperfectly and surreptitiously published before: In 1642 Andrew Crooke had published an edition of *Religio Medici* without Browne's permission.

4 *dissentaneous:* Disagreeing with, dissenting from.

Tropicall: Figurative or highly rhetorical expressions (the use of tropes).

5 *scandall of my profession:* The medical profession in the seventeenth century was often associated with atheism.

6 *decaied bottome:* Vessel unfit for seafaring (Browne is describing Roman Catholics).

improperations: Insults.

difference our affections: Distinguish our attitudes, practices, and rituals (but not "our cause," that is, Christianity).

8 *difference my self neerer:* Be more specific about my religion.

points indifferent: Non-doctrinal matters.

Councell of Trent: Several ecumenical councils between 1545 and 1563 determining the platform of the Catholic Counter-Reformation.

Synod of Dort: Synod of Reformed Churches in 1618 to settle the major points of orthodox Protestantism.

9 *his best Oedipus:* That is, for solving riddles.

10 *Metempsuchosis:* The transmigration of the soul from one body to another after death.

reflex: Reflection, understanding.

11 *Arians:* Those who believe that the Son is not equal to or of the same substance as God the Father (a position defined as heretical by early ecumenical councils).

12 *complexionally propense:* Naturally inclined.

Pia Mater: The innermost membrane surrounding the brain and spinal cord.

o altitudo: An expression of the unfathomable mysteries of God's wisdom and power (see Romans 11:33: "O the depth of the riches and wisdom and knowledge of God! How unsearchable are his judgements and how inscrutable his ways").

Certum est quia impossibile est: It is true (or certain) because it is impossible.

13 *mysticall Types:* Figures in the Old Testament who resemble and prefigure Christ.

Buckler: Shield (see Ephesians 6:16: "With all of these, take the shield of faith, with which you will be able to quench all the flaming arrows of the evil one. Take the helmet of salvation, and the sword of the Spirit, which is the word of God").

Deus est Sphæra cuius centrum ubique. . . : "God is a sphere whose center is everywhere and whose circumference is nowhere."

I had as lieve you tell me . . . : Browne prefers Christian tropes (the soul is man's angel and the body of God, light is the shadow of God) to what he sees as the jargon of scholastic philosophy.

haggard: Untamed (the imagery is drawn from the practice of falconry).

Pucellage: Maidenhood or virginity.

14 *Neque enim cum porticus aut me lectulus accepit, desum mihi:* "Nor do I forget myself when retired to my couch or portico" (Browne is slightly emending Horace's *Satire* I, iv, 133–34).

14 *Saint Paul's Sanctuary:* The recognition that God's ways are beyond rational understanding or definition.

prescious: Foreknowing.

placet: Something committed or assented to.

15 *Trinitie of our soules:* Aristotle describes the soul in terms of its distinct vegetative, sensitive, and rational capacities.

front: Forehead.

roundles: Rungs.

16 *γνω῀ϑισεαυτόν nosce teipsum:* "Know thyself," in Greek and Latin respectively. In the text above, Browne expresses the common early-modern belief that ancient oracles were the work of the devil.

consultation and election: Deliberation and choice.

to profound: Used as a verb meaning "to explore the depths of."

Sanctum sanctorum: Holiest of holies, the innermost place in a temple or tabernacle.

17 *foure second causes:* Aristotle's material, formal, efficient, and final causes; for example the cause of a house is at once its bricks and wood (material), its plan or blueprint (formal), the work of the builders (efficient), and its intended use as a shelter (final).

18 *Galen . . . Suarez:* Browne claims that an appreciation of God's ways can be as easily gained from the physiological treatises of Galen (2nd century CE) as from the Spanish theologian Francisco Suarez (1548–1617).

this cause: The final cause.

Natura nihil agit frustra: "Nature does nothing in vain" (one of Aristotle's most significant assumptions).

cantons: Spaces.

Regio-Montanus: Fifteenth-century mathematician and astronomer, supposed to have invented both a mechanical fly and eagle.

19 *its supernaturall station:* See Joshua 10:13: "And the sun stood still, and the moon stopped, until the nation took vengeance on their enemies."

19 *sweetned the water with a wood:* See Exodus 15:24–25: "And the people complained against Moses, saying, 'What shall we drink?' He cried out to the Lord; and the Lord showed him a piece of wood; he threw it into the water, and the water became sweet."

20 *Ephemerides:* A book tabulating daily planetary positions, frequently used for horoscopy.

21 *Bezo las Manos:* A salutation by kissing of the hands.

Gramercy: Thank you.

the Ram in the thicket: See Genesis 22:13: "And Abraham looked up and saw a ram, caught in a thicket by its horns. Abraham went and took the ram and offered it up as a burnt-offering instead of his son."

Moses in the Arke: See Exodus 2:5: "The daughter of Pharaoh came down to bathe at the river, while her attendants walked beside the river. She saw the basket among the reeds and sent her maid to bring it."

the story of Joseph: See Genesis 37–50.

rubs, doublings and wrenches: Difficulties, coincidences, and surprises, all seemingly chance occurrences.

Fougade or Powder Plot: The famous Gunpowder Plot of 1605 was uncovered by a letter sent to Lord Monteagle.

intelligences: Spirits or angels thought to control the motion of planets.

22 *Helix that still enlargeth:* An ever expanding spiral.

sortilegies: A form of divination, generally by casting lots.

donatives: Gifts.

engrosse: Arrogate, claim the entirety of something.

23 *judiciall Astrology:* Branches of astrology (e.g., constructing a natal chart to predict the future) deemed heretical.

23 *supputation:* Accounting.

23 *Homers chaine:* The chain connecting the heavens and earth in Homer's *Iliad.*

Sorites: A chain of syllogisms (more generally an intricate argument).

advisoes: Counsel, guidance.

24 *Archidoxis:* A treatise by Paracelsus on talismanic or sympathetic magic, that is, magic based on imitation or correspondence.

the Brazen Serpent: See Numbers 21:9: "So Moses made a serpent of bronze, and put it upon a pole; and whenever a serpent bit someone, that person would look at the serpent of bronze and live."

a miracle in Elias: See I Kings 18:30–35.

25 *Moses, Christ, and Mahomet: The Treatise of the Three Impostors* was a legendary book denouncing Moses, Jesus, and Muhammad; attributed to various writers in medieval and early-modern Europe.

prejudicate beleefe: An opinion not subjected to rational scrutiny.

gravelled: Disturbed.

Post mortem nihil est . . . : From Seneca's *Troades.* "There is nothing after death, and death is itself nothing. Death is indivisible, destroying the body and soul together. We die completely, and no part of us remains."

26 *Garagantua or Bevis:* Gargantua, a giant from the sixteenth-century work of François Rabelais; Bevis was a knight of medieval romance.

edified: Created, composed.

27 *Pantagruel:* Another of Rabelais's giants, in whose library one finds the mentioned treatise *De Modo Cacandi*, or "On the Method of Shitting."

Deucalion: In Greek mythology, Deucalion and his wife, Pyrrha, were the only survivors of a global deluge, similar to the one recounted in Genesis.

particular: Relegated to a particular region or country, rather than a global flood.

28 *compellations:* Ways of addressing each other (as between equals).

Tutelary Angels: Guardian angels.

29 *This without a blow:* The Bible (by way of contrast with the Koran).

Enochs Pillars: Enoch was said to have inscribed on two pillars all the great events, arts, and wisdom of the antediluvian world.

30 *three great inventions of Germany:* Gunpowder, the printing press, and the mariner's compass.

Utinam: Latin exclamation (If only! Would that . . . !).

Ethnick: Pagan.

32 *Councell of Constance:* Ecumenical council of 1414 to 1418.

Virgilius: Eighth-century bishop of Salzburg who came into conflict with church authorities for arguing in favor of the earth's rotundity ("antipodes" are any two points diametrically opposite each other on the earth's surface).

33 *Esdras:* See II Esdras 4:5: "I said, 'Speak on, my lord.' And he said to me, 'Go, weigh for me the weight of fire, or measure for me a measure of wind, or call back for me the day that is past.'"

the efficacy of reliques: Browne dismisses the Roman Catholic fervor for holy relics, and cites three of the most famous: Helena claimed to have found the true cross; Constantine was said to have worn a nail from the Crucifixion in his helmet; and Baldwin claimed to have found the ashes of John the Baptist.

Piae fraudes: Holy frauds.

34 *Climacter:* Climacteric: any critical stage of a person's life marking major changes in health or fortune.

cessation of Oracles: It was believed that the pagan oracles ceased to prophecy after the birth of Christ.

Hester: Esther.

35 *detection:* Revelation of fraud.

35 *upon the smell of a Rose:* Eva Flegen, a famous spiritual impostor, was rumored to have fasted for years, surviving only on the smell of a rose.

36 *Ascendens constellatum . . . :* A rising star reveals many thing to those who seek the wonders of nature, that is, the works of God.

Hermeticall Philosophers: Philosophers who, following the teachings of Hermes Trismegistus, believe in a spirit world corresponding to the earthly one, accessible by alchemy and astrology.

38 *specificall difference:* Angels have immediate knowledge of innate differences between species, whereas humans understand less certainly by observation of externals ("accidents and properties").

reserved difference: Special distinction of each individual within a given species.

Hypostasis: Individual substance.

39 *Habakkuk to the Lions den:* See the narrative of Bel and the Dragon, included in a portion, sometimes considered apocryphal, of *The Book of Daniel* (1:36).

Philip to Azotus: See Acts 8:39–40: "When they came up out of the water, the Spirit of the Lord snatched Philip away; the eunuch saw him no more, and went on his way rejoicing. But Philip found himself at Azotus, and as he was passing through the region, he proclaimed the good news to all the towns until he came to Caesarea."

great Father: Saint Augustine.

Fiat lux: Let there be light.

preferred: Promoted.

40 *of the other:* The visible world, which Moses described only obscurely so that some parts of it (like the element of fire supposed to surround the earth) remained uncertain in Browne's time.

first moveable: Primum mobile: first part of the cosmos set in motion according to the Ptolemaic system.

41 *Omneity informed Nullity:* The spirit of All (God) informed Nothingness.

For these two affections: The soul's incorruptibility and immortality are affirmed by Plato and at least not denied by Aristotle.

auditories: Auditoriums or learned assemblies.

without conjunction: Browne refers to the theory of Paracelsus that the human being in miniature was fully formed and preexistent in the sperms cells, allowing for a human existence before any act of copulation.

traduction: The belief (which Browne held) that the soul is passed naturally and materially in the process of procreation from parent to child, and against which he cites Augustine's description of the soul's infusion into the body: "In creation it is infused, in being infused it is created."

A figure in Rethoricke . . . : Augustine's chiasmus (above) is an example of antimetathesis.

42 *Crasis:* Composition or mixture.

carnified: Embodied, made flesh.

43 *Lots wife:* See Genesis 19:26: "But Lot's wife, behind him, looked back, and she became a pillar of salt."

Nabuchodonosor: See Daniel 4:33: "Immediately the sentence was fulfilled against Nebuchadnezzar. He was driven away from human society, ate grass like oxen, and his body was bathed with the dew of heaven, until his hair grew as long as eagles' feathers and his nails became like birds' claws."

Adam, quid fecisti?: See Genesis 3:13: "Adam, what have you done?"

Vespilloes: One who carried out the bodies of the poor at night for burial.

45 *secondine:* Protective layer surrounding the fetus.

ubi: Place.

perfect exaltation: Highest refinement.

45 *est mutatio . . . :* "It is the change that perfects that noble part of the microcosm."

46 *quantum mutatus ab illo:* "How changed from that (which he was)," spoken of Hector in *Aeneid* (2: 274).

Rodomontado: Any sort of blustering or bragging speech.

47 *Canicular dayes:* Dog days, youth.

incurvate: Bend.

48 *vitiosity:* Sinfulness.

Aesons bath: By means of a magical bath Medea restored the youth of Jason's father, Aeson.

radicall humour: One of the ultimate material elements of living beings.

radicall balsome or vitall sulphur: Considered by alchemists as one of the vital elements of all material forms of life.

glome: Thread or clue.

49 *Emori nolo . . . :* From Cicero's *Tusculan Disputations* (I, viii).

50 *Memento quatuor novissima:* "Remember the four last things."

51 *Lucan:* From *Pharsalia* (VII, 814–15).

convincible: Deserving of punishment.

Elias 6000 yeares: The belief (attributed to the writings of Elijah) that the world would last for 6,000 years.

quære: Question, query.

52 *amphibology:* An ambiguous figure of speech.

respective: Unequal.

53 *resolution of his:* Seneca's.

54 *mortified:* Separated or dissolved.

sensible Artist: A practitioner (like Browne) endowed with a powerful observation and imagination.

55 *types:* Anticipations or foreshadowings.

55 *That elegant Apostle:* Saint Paul.

Empyreall: In the Ptolemaic system, the tenth stationary sphere beyond the mobile spheres of the planets and stars.

56 *the Parable:* See Luke 16:19–31.

glorified creatures: The state of existence after the resurrection of the body.

Perspective: Telescope.

visible species: Appearances transmitted by external objects.

textuaries: Specialists in the study of Scripture.

57 *reverberated:* Transformed.

in posse: Existing only potentially.

58 *Legion:* The demonic spirits of Mark 5:9: "Then Jesus asked him, 'What is your name?' He replied, 'My name is Legion; for we are many.'"

conceited: Postulated.

impassible: Not subject to pain or harm.

60 *controvert:* Deny the truth of.

perpend: Consider or examine thoroughly.

Simile of Saint Paul: See Romans 9:20–21.

Phalaris his Bull: Infamous instrument of torture in which victims were roasted alive.

stone: Kidney stones.

61 *profound:* Plunge or sink.

venue: Arrival.

under heads: The less refined, the vulgar.

Chiron: Mythical centaur, tutor to Achilles.

Strabo's cloake: The ancient geographer Strabo described the shape of the known world as resembling a tunic or cloak.

62 *Sectaries:* Dissenting Protestants.

Atomist: Materialist, one who believes the soul dies with the body.

Familist: Member of the Family of Love, a Protestant sect in England from the reign of Elizabeth.

63 *the precept of Saint Paul:* See Philippians 2:12.

beneplacit: Pleasure.

the saying of Christ: See John 8:58.

64 *Midianites:* See Judges 7:4–8.

object unto: Reveal or place in front of.

65 *eighth Climate:* England ("climate" here is a measure of latitude).

66 *Mechanickes:* Laborers in general.

Doradoes: Golden ones, men of wealth.

67 *Eleemosynaries:* Those who live on charity.

68 *a la volee:* Without forethought or design.

Chiromancy: Palmistry.

vagabond and counterfeit Egyptians: Gypsies.

its Copy: What it copies, the Platonic idea or pattern in the mind of God.

69 *caitif:* Base or conwardly.

70 *βατραχομυομαχία:* A mock-epic battle between mice and frogs, once thought to have been the work of Homer.

S. and T. in Lucian: A comic trial between the Greek letters sigma and tau.

Priscian: Roman grammarian; "to break Priscian's head" was a common expression applied to those who spoke or wrote Latin poorly.

Si foret in terris . . . : "If Democritus were still on earth he would laugh" (from Horace's *Espistle* II, i, 194).

70 *Basilisco:* A piece of heavy artillery.

71 *Le mutin Anglois . . . :* "The rebellious English, the swaggering Scot, / the bugger Italian, the crazy French, / the cowardly Roman, the thieving Gascon, / the vain Spaniard, the drunken German" (Browne is adapting Joachim du Bellay's *Les Regrets*).

Saint Paul: See Titus 1:12.

as Neroes was in another: Browne is probably thinking of Caligula, who wished all Romans had a single neck to cut.

72 *derived ray:* Indirect or reflected image.

trajection of sensible species: The transmission of the material image of an object to the mind.

quadrate: To agree with.

Non occides: Thou shalt not kill.

73 *like a dimension:* Divisible like any object with dimensions.

Nisus . . . Patroclus: Classical examples of great friendship. In Greek mythology, Damon offers his life as bond for his friend Pythias. The story of Nisus and Euryalus is given in Virgil's *Aeneid*, while that of Achilles and Patroclus is found in Homer's *Iliad*.

the fifth Commandement: Honor thy father and mother.

74 *three most mysticall unions:* These are the two natures of Christ (human and divine), the Trinity, and the union of two souls through friendship.

75 *the story of the Italian:* "Who, after he had inveigled his enemy to disclaim his faith for the redemption of his life, did presently poniard him to prevent repentance and assure his eternal death," a story Browne relates in his *Pseudodoxia Epidemica* (VII, 19).

battell of Lepanto: Decisive naval victory of combined European forces against the fleet of the Ottoman Empire on October 7, 1571.

75 *dastards mee:* Renders me dishonorable or weak.

76 *humorous:* Based on the ancient medical theory of the four humors.

76 *Nero in his Spintrian recreations:* Browne is actually recalling Emperor Tiberius and his sexually perverse practices.

77 *Chorography:* The mapping or geographical description of a region.

Poynters: The two stars of the Great Bear.

Simpled: Gathered herbs for medicinal purposes.

Cheap-side: One of the major thoroughfares of London.

riddle of the Fishermen: A legend according to which Homer pined away and died after being unable to solve a riddle posed by a fisherman.

Euripus: In his *Pseudodoxia Epidemica* (VII, 13), Browne denies the popular legend that "Aristotle drowned himself in Euripus as despairing to resolve the cause of its reciprocation, or ebb and flow seven times a day."

78 *Peripateticks:* Followers of Aristotle.

Academicks: Followers of Plato.

glorification: The resurrection of the body.

80 *Physicke:* Medicine.

Catholicon: Panacea.

81 *Magnæ virtutes . . . :* Great were his virtues, and no less were his vices.

Antiperistasis: An opposition that strengthens the quality opposed.

Nunquam minus solus quam cum solus: Never less alone than when alone.

82 *Ruat coelum, Fiat voluntas tua:* "Though the heavens fall, let thy will be done."

83 *stiver:* A Dutch coin of small value.

ligation: Tying up.

a peece of that Leaden Planet: A melancholy disposition.

galliardize: Excessive gaiety.

84 *Themistocles:* An Athenian statesman and general. However, Browne is most likely recalling an anecdote about another Greek general, Iphicrates, who slew a soldier for falling asleep during the night watch.

Lucan and Seneca: Both of whom were allowed by Nero to choose the manner of their death.

85 *supererogate in that common principle:* To do more for another than is required.

86 *subterraneous Idoll:* Gold.

the example of the Mite: See Mark 12:41–44: "[Jesus] sat down opposite the treasury, and watched the crowd putting money into the treasury. Many rich people put in large sums. A poor widow came and put in two small copper coins, which are worth a penny. Then he called his disciples and said to them, 'Truly I tell you, this poor widow has put in more than all those who are contributing to the treasury. For all of them have contributed out of their abundance; but she out of her poverty has put in everything she had, all she had to live on.'"

Peru: Famous for its riches.

87 *Centoes:* Poems composed entirely of lines from other poems. Browne uses the term metaphorically for the patchwork clothing of the poor.

traduction: Transmission.

call to assize: Examine.

88 *Crambe:* Tiresome repetition.

summum bonum: Highest good.

a story out of Pliny: A fabulous or unbelievable narration.

HYDRIOTAPHIA, OR URNE-BURIALL

97 *challenge a restitution:* Claim a right to be returned (to the earth).

yet few have returned their bones farre lower than they might receive them: Few have left their bones far lower than the surface of the earth, which formed and sustained them.

centrall interrment: Deep burial (near the earth's center).

some parts, which they never beheld themselves: Their own bones.

98 *the hot contest between Satan and the Arch-Angell, about discovering the body of Moses:* See Jude 1:9: "But when the archangel Michael contended with the devil and disputed about the body of Moses, he did not dare to bring a condemnation of slander against him, but said, 'The Lord rebuke you!'"

as low as: As late as.

99 *Not totally pursued in the highest runne of Cremation:* Cremation, though increasingly popular, was not exclusively practiced.

conclude in a moist relentment: Believe that our bodies return to water.

100 *waved the fiery solution:* Refused cremation.

affected: Chose or preferred.

101 *affecting rather a depositure than absumption:* Choosing interment over cremation.

For the men of Jabesh burnt the body of Saul: See 1 Samuel 31:12.

102 *the expressions concerning Jehoram, Sedechias, and the sumptuous pyre of Asa:* For the fires on the occasions of their death, see 2 Chronicles 21:19, Jeremiah 34:5, and 2 Chronicles 16:14.

not scrupulous in: Did not hesitate to.

hottest use: Greatest abuse.

the types of Enoch, Eliah, or Jonah: Figures interpreted by Christians as anticipating or foreshadowing the Resurrection.

fasciations: Bonds.

103 *Civilians . . . animals:* Those who study civil law or custom (especially Roman) assume that interment is a strictly human practice, while others find examples of it in the animal kingdom.

104 *Ustrina:* A sacred enclosure in which Romans would cremate their dead.

Manes: Spirits of the dead according to Roman mythology.

Aræ: Roman altar.

105 *Emphaticall appellation:* The true sense of the name.

106 *leave to higher conjecture:* Leave to the expert antiquarians to decide.

abridgeth such Antiquity: Implies that it is not as old as the time of Caesar.

107 *Lacrymatories:* A bottle or small vase used for holding tears.

will challenge above: Will claim an antiquity of more than 1,300 years.

obviously objected upon Christians: Christians were held in suspicion because of their disapproval of cremation.

108 *exility:* Smallness.

109 *Septuagint:* The ancient Greek translation of the Hebrew Bible.

frugall Bit: Small piece of food.

110 *raised an Æra:* Demarcated a historical period based on the institution of cremation.

111 *Ulysses*: The character Odysseus in Euripides' play *Hecuba*.

112 *making our last bed like our first:* The shape of some of the urns resembles the womb.

Microcosme: Referring to the popular Renaissance belief that the human being was structured like a little world.

testaceous: Having a hard, shell-like outer covering.

testa: A kind of earthen pot.

113 *Tegument:* A roof or covering.

113 *incrassated:* Made hard, thick, or solid.

farre exceeded the Palats of Antiquity: Tasted a wine so long aged that nothing in the ancient world could compare to it.

Consulary date: Roman wine was stamped with the date of the current consul.

Opimian Wine: A famous Roman vintage under the consulate of Lucius Opimius (121 BCE).

114 *conservatories:* Any places where things were kept or stored for preservation.

vegetable Antiquity: Any naturally growing substance (for example, wood) having been preserved for great lengths of time.

we admire not: We are not surprised at or incredulous of.

115 *designed:* Designated.

confounded: Mixed together.

116 *Without confused burnings they affectionately compounded their bones:* Though not burned together, their bones were afterward mixed or placed in the same urn.

so curious to: So intent on, determined to.

117 *it is obvious to meet with:* Inevitably we find.

patera's: Roman dishes used for drinking libations and in other ritual contexts.

Hypogæum: Vault.

iterately affecting: Frequently choosing.

Pismires: Ants.

Election of such differences: Choice between alternative accounts.

Cenotaphs: Empty tombs erected to honor those whose remains lie elsewhere.

118 *Expilators:* Plunderers.

118 *terra damnata:* An alchemical term for the residue left after sublimation or calcination.

Plato's historian of the other world: The figure Er from *Republic*, Book X.

exenteration: The surgical removal of the inner organs.

fictile Vessels: Vessels made of earth or clay.

119 *incremable:* Unable to be burned.

ponderation: Weight, heaviness.

120 *Amos 2:1*: "Thus says the Lord: For three transgressions of Moab and for four, I will not revoke the punishment; because he burned to lime the bones of the king of Edom."

ferity: The condition of being savage; ferocity.

tests: Vessels or outer shells.

121 *promiscuous practise:* Frequent or customary use.

To be gnaw'd out of our graves: Rudely unearthed (NB: "gnaw'd" is a correction for "knav'd," but some editors believe Browne intended the latter).

122 *fat concretion:* Adipocere, a waxlike organic substance formed by the hydrolysis of body fats. Browne is credited with being the first to note its existence and describe its nature.

lixivious: Alkaline.

the opprobrious disease: Syphilis.

Arefaction: The act of drying.

123 *identifie:* The divine power to re-create identity by recomposing the scattered parts of a creature.

124 *they devolved not all upon the sufficiency of soul existence:* They did not rely on the immortality of the soul to the extent of ignoring funerary rites.

the Greek devotion: The rituals of the Eastern Orthodox Church.

124 *Demas and Soma:* Homer's distinction between the living (*Demas*) and dead (*Soma*) body.

125 *transcorporating Philosophers:* Philosophers (such as Pythagoras) who upheld metempyschosis, or the transmigration of souls.

126 *exuccous:* Dry or sapless.

Mahometans: Muslims.

127 *iterated clamations:* Repeated cries.

facilitating the accension: The kindling of the pyre.

128 *Nor were only many customes questionable . . . :* Here and in the following paragraphs Browne cites contradictory accounts of the afterlife among such writers as Homer, Lucian, Virgil, and Dante.

Morta: The deity of death or fate.

130 *ancient Martyrs:* Elderly martyrs (as opposed to martyrs from ancient times).

corporall animosity: Courage, spiritedness.

the immortality of Plato: Plato's theory of the soul's natural immortality.

131 *diuturnity:* Long duration.

Sic ego componi versus in ossa velim: "So I would wish to be buried when I turn to bones" (Tibullus III, ii, 26).

132 *propension:* Propensity, inclination.

133 *admit a wide solution:* Allow for a wide range of conjectures.

tutellary Observators: Guardian angels.

probable Meridian of time: Midpoint of human history.

we cannot expect such Mummies unto our memories: We cannot hope for our memories to last as long as those of the ancient heroes.

135 *Entelechia:* The essence and actualization of the soul, according to Aristotle.

135 *The Canaanitish woman:* See Matthew 5:21–28.

the everlasting Register: The Holy Scriptures.

makes but winter arches: Provides the light of only a winter's day.

136 *callosities:* Hardness, callousness.

Mummie is become Merchandise: In Browne's time "mummy" was also a drug obtained from mummified human remains.

137 *perspectives:* Telescopes, such as Galileo had used to discover spots in the sun, disproved Aristotle's idea that the region above the moon was not subject to change or decay.

138 *decretory term of the world:* The Day of Judgment.

139 *exolution:* The soul's departure from the body.

prædicament of Chymera's: Vain creations of the imagination.

Tabesne cadavera solvat / An rogus haud refert: "Nor does it matter whether the corpses decompose or are burnt on the pyre" (Lucan, *Pharsalia*, VII, 809–10).

TITLES IN SERIES

For a complete list of titles, visit www.nyrb.com or write to:
Catalog Requests, NYRB, 435 Hudson Street, New York, NY 10014

J.R. ACKERLEY Hindoo Holiday*
J.R. ACKERLEY My Dog Tulip*
J.R. ACKERLEY My Father and Myself*
J.R. ACKERLEY We Think the World of You*
HENRY ADAMS The Jeffersonian Transformation
CÉLESTE ALBARET Monsieur Proust
DANTE ALIGHIERI The Inferno
DANTE ALIGHIERI The New Life
WILLIAM ATTAWAY Blood on the Forge
W.H. AUDEN (EDITOR) The Living Thoughts of Kierkegaard
W.H. AUDEN W.H. Auden's Book of Light Verse
ERICH AUERBACH Dante: Poet of the Secular World
DOROTHY BAKER Cassandra at the Wedding
J.A. BAKER The Peregrine
HONORÉ DE BALZAC The Unknown Masterpiece *and* Gambara*
MAX BEERBOHM Seven Men
STEPHEN BENATAR Wish Her Safe at Home*
FRANS G. BENGTSSON The Long Ships*
ALEXANDER BERKMAN Prison Memoirs of an Anarchist
GEORGES BERNANOS Mouchette
ADOLFO BIOY CASARES Asleep in the Sun
ADOLFO BIOY CASARES The Invention of Morel
CAROLINE BLACKWOOD Corrigan*
CAROLINE BLACKWOOD Great Granny Webster*
NICOLAS BOUVIER The Way of the World
MALCOLM BRALY On the Yard*
MILLEN BRAND The Outward Room*
SIR THOMAS BROWNE Religio Medici *and* Urne-Buriall*
JOHN HORNE BURNS The Gallery
ROBERT BURTON The Anatomy of Melancholy
CAMARA LAYE The Radiance of the King
GIROLAMO CARDANO The Book of My Life
DON CARPENTER Hard Rain Falling*
J.L. CARR A Month in the Country
BLAISE CENDRARS Moravagine
EILEEN CHANG Love in a Fallen City
UPAMANYU CHATTERJEE English, August: An Indian Story
NIRAD C. CHAUDHURI The Autobiography of an Unknown Indian
ANTON CHEKHOV Peasants and Other Stories
RICHARD COBB Paris and Elsewhere
COLETTE The Pure and the Impure
JOHN COLLIER Fancies and Goodnights
CARLO COLLODI The Adventures of Pinocchio*
IVY COMPTON-BURNETT A House and Its Head
IVY COMPTON-BURNETT Manservant and Maidservant
BARBARA COMYNS The Vet's Daughter
EVAN S. CONNELL The Diary of a Rapist
ALBERT COSSERY The Jokers*

* *Also available as an electronic book.*

ALBERT COSSERY Proud Beggars*
HAROLD CRUSE The Crisis of the Negro Intellectual
ASTOLPHE DE CUSTINE Letters from Russia
LORENZO DA PONTE Memoirs
ELIZABETH DAVID A Book of Mediterranean Food
ELIZABETH DAVID Summer Cooking
L.J. DAVIS A Meaningful Life*
VIVANT DENON No Tomorrow/Point de lendemain
MARIA DERMOÛT The Ten Thousand Things
DER NISTER The Family Mashber
TIBOR DÉRY Niki: The Story of a Dog
ARTHUR CONAN DOYLE The Exploits and Adventures of Brigadier Gerard
CHARLES DUFF A Handbook on Hanging
BRUCE DUFFY The World As I Found It*
DAPHNE DU MAURIER Don't Look Now: Stories
ELAINE DUNDY The Dud Avocado*
ELAINE DUNDY The Old Man and Me*
G.B. EDWARDS The Book of Ebenezer Le Page
MARCELLUS EMANTS A Posthumous Confession
EURIPIDES Grief Lessons: Four Plays; translated by Anne Carson
J.G. FARRELL Troubles*
J.G. FARRELL The Siege of Krishnapur*
J.G. FARRELL The Singapore Grip*
ELIZA FAY Original Letters from India
KENNETH FEARING The Big Clock
KENNETH FEARING Clark Gifford's Body
FÉLIX FÉNÉON Novels in Three Lines*
M.I. FINLEY The World of Odysseus
THEODOR FONTANE Irretrievable*
EDWIN FRANK (EDITOR) Unknown Masterpieces
MASANOBU FUKUOKA The One-Straw Revolution*
MARC FUMAROLI When the World Spoke French
CARLO EMILIO GADDA That Awful Mess on the Via Merulana
MAVIS GALLANT The Cost of Living: Early and Uncollected Stories*
MAVIS GALLANT Paris Stories*
MAVIS GALLANT Varieties of Exile*
GABRIEL GARCÍA MÁRQUEZ Clandestine in Chile: The Adventures of Miguel Littín
ALAN GARNER Red Shift*
THÉOPHILE GAUTIER My Fantoms
JEAN GENET Prisoner of Love
ÉLISABETH GILLE The Mirador: Dreamed Memories of Irène Némirovsky by Her Daughter*
JOHN GLASSCO Memoirs of Montparnasse*
P.V. GLOB The Bog People: Iron-Age Man Preserved
NIKOLAI GOGOL Dead Souls*
EDMOND AND JULES DE GONCOURT Pages from the Goncourt Journals
EDWARD GOREY (EDITOR) The Haunted Looking Glass
A.C. GRAHAM Poems of the Late T'ang
WILLIAM LINDSAY GRESHAM Nightmare Alley*
EMMETT GROGAN Ringolevio: A Life Played for Keeps
VASILY GROSSMAN Everything Flows*
VASILY GROSSMAN Life and Fate*
VASILY GROSSMAN The Road*
OAKLEY HALL Warlock

PATRICK HAMILTON The Slaves of Solitude
PATRICK HAMILTON Twenty Thousand Streets Under the Sky
PETER HANDKE Short Letter, Long Farewell
PETER HANDKE Slow Homecoming
ELIZABETH HARDWICK The New York Stories of Elizabeth Hardwick*
ELIZABETH HARDWICK Seduction and Betrayal*
ELIZABETH HARDWICK Sleepless Nights*
L.P. HARTLEY Eustace and Hilda: A Trilogy*
L.P. HARTLEY The Go-Between*
NATHANIEL HAWTHORNE Twenty Days with Julian & Little Bunny by Papa
GILBERT HIGHET Poets in a Landscape
JANET HOBHOUSE The Furies
HUGO VON HOFMANNSTHAL The Lord Chandos Letter
JAMES HOGG The Private Memoirs and Confessions of a Justified Sinner
RICHARD HOLMES Shelley: The Pursuit
ALISTAIR HORNE A Savage War of Peace: Algeria 1954–1962*
GEOFFREY HOUSEHOLD Rogue Male*
WILLIAM DEAN HOWELLS Indian Summer
BOHUMIL HRABAL Dancing Lessons for the Advanced in Age*
DOROTHY B. HUGHES The Expendable Man*
RICHARD HUGHES A High Wind in Jamaica*
RICHARD HUGHES In Hazard
RICHARD HUGHES The Fox in the Attic (The Human Predicament, Vol. 1)
RICHARD HUGHES The Wooden Shepherdess (The Human Predicament, Vol. 2)
MAUDE HUTCHINS Victorine
YASUSHI INOUE Tun-huang*
HENRY JAMES The Ivory Tower
HENRY JAMES The New York Stories of Henry James*
HENRY JAMES The Other House
HENRY JAMES The Outcry
TOVE JANSSON Fair Play
TOVE JANSSON The Summer Book
TOVE JANSSON The True Deceiver
RANDALL JARRELL (EDITOR) Randall Jarrell's Book of Stories
DAVID JONES In Parenthesis
KABIR Songs of Kabir; translated by Arvind Krishna Mehrotra*
FRIGYES KARINTHY A Journey Round My Skull
HELEN KELLER The World I Live In
YASHAR KEMAL Memed, My Hawk
YASHAR KEMAL They Burn the Thistles
MURRAY KEMPTON Part of Our Time: Some Ruins and Monuments of the Thirties
RAYMOND KENNEDY Ride a Cockhorse*
DAVID KIDD Peking Story*
ROBERT KIRK The Secret Commonwealth of Elves, Fauns, and Fairies
ARUN KOLATKAR Jejuri
DEZSŐ KOSZTOLÁNYI Skylark*
TÉTÉ-MICHEL KPOMASSIE An African in Greenland
GYULA KRÚDY The Adventures of Sindbad*
GYULA KRÚDY Sunflower*
SIGIZMUND KRZHIZHANOVSKY The Letter Killers Club*
SIGIZMUND KRZHIZHANOVSKY Memories of the Future
MARGARET LEECH Reveille in Washington: 1860–1865*
PATRICK LEIGH FERMOR Between the Woods and the Water*

PATRICK LEIGH FERMOR Mani: Travels in the Southern Peloponnese*
PATRICK LEIGH FERMOR Roumeli: Travels in Northern Greece*
PATRICK LEIGH FERMOR A Time of Gifts*
PATRICK LEIGH FERMOR A Time to Keep Silence*
PATRICK LEIGH FERMOR The Traveller's Tree*
D.B. WYNDHAM LEWIS AND CHARLES LEE (EDITORS) The Stuffed Owl
GEORG CHRISTOPH LICHTENBERG The Waste Books
JAKOV LIND Soul of Wood and Other Stories
H.P. LOVECRAFT AND OTHERS The Colour Out of Space
DWIGHT MACDONALD Masscult and Midcult: Essays Against the American Grain*
NORMAN MAILER Miami and the Siege of Chicago*
JANET MALCOLM In the Freud Archives
JEAN-PATRICK MANCHETTE Fatale*
OSIP MANDELSTAM The Selected Poems of Osip Mandelstam
OLIVIA MANNING Fortunes of War: The Balkan Trilogy
OLIVIA MANNING School for Love*
JAMES VANCE MARSHALL Walkabout*
GUY DE MAUPASSANT Afloat
GUY DE MAUPASSANT Alien Hearts*
JAMES MCCOURT Mawrdew Czgowchwz*
HENRI MICHAUX Miserable Miracle
JESSICA MITFORD Hons and Rebels
JESSICA MITFORD Poison Penmanship*
NANCY MITFORD Madame de Pompadour*
NANCY MITFORD The Sun King*
HENRY DE MONTHERLANT Chaos and Night
BRIAN MOORE The Lonely Passion of Judith Hearne*
BRIAN MOORE The Mangan Inheritance*
ALBERTO MORAVIA Boredom*
ALBERTO MORAVIA Contempt*
JAN MORRIS Conundrum
JAN MORRIS Hav*
PENELOPE MORTIMER The Pumpkin Eater*
ÁLVARO MUTIS The Adventures and Misadventures of Maqroll
L.H. MYERS The Root and the Flower*
NESCIO Amsterdam Stories*
DARCY O'BRIEN A Way of Life, Like Any Other
YURI OLESHA Envy*
IONA AND PETER OPIE The Lore and Language of Schoolchildren
IRIS OWENS After Claude*
RUSSELL PAGE The Education of a Gardener
ALEXANDROS PAPADIAMANTIS The Murderess
BORIS PASTERNAK, MARINA TSVETAYEVA, AND RAINER MARIA RILKE Letters, Summer 1926
CESARE PAVESE The Moon and the Bonfires
CESARE PAVESE The Selected Works of Cesare Pavese
LUIGI PIRANDELLO The Late Mattia Pascal
ANDREY PLATONOV The Foundation Pit
ANDREY PLATONOV Soul and Other Stories
J.F. POWERS Morte d'Urban
J.F. POWERS The Stories of J.F. Powers
J.F. POWERS Wheat That Springeth Green
CHRISTOPHER PRIEST Inverted World
BOLESŁAW PRUS The Doll*

RAYMOND QUENEAU We Always Treat Women Too Well
RAYMOND QUENEAU Witch Grass
RAYMOND RADIGUET Count d'Orgel's Ball
JULES RENARD Nature Stories*
JEAN RENOIR Renoir, My Father
GREGOR VON REZZORI An Ermine in Czernopol*
GREGOR VON REZZORI Memoirs of an Anti-Semite*
GREGOR VON REZZORI The Snows of Yesteryear: Portraits for an Autobiography*
TIM ROBINSON Stones of Aran: Labyrinth
TIM ROBINSON Stones of Aran: Pilgrimage
MILTON ROKEACH The Three Christs of Ypsilanti*
FR. ROLFE Hadrian the Seventh
GILLIAN ROSE Love's Work
WILLIAM ROUGHEAD Classic Crimes
CONSTANCE ROURKE American Humor: A Study of the National Character
TAYEB SALIH Season of Migration to the North
TAYEB SALIH The Wedding of Zein*
GERSHOM SCHOLEM Walter Benjamin: The Story of a Friendship
DANIEL PAUL SCHREBER Memoirs of My Nervous Illness
JAMES SCHUYLER Alfred and Guinevere
JAMES SCHUYLER What's for Dinner?*
LEONARDO SCIASCIA The Day of the Owl
LEONARDO SCIASCIA Equal Danger
LEONARDO SCIASCIA The Moro Affair
LEONARDO SCIASCIA To Each His Own
LEONARDO SCIASCIA The Wine-Dark Sea
VICTOR SEGALEN René Leys
PHILIPE-PAUL DE SÉGUR Defeat: Napoleon's Russian Campaign
VICTOR SERGE The Case of Comrade Tulayev*
VICTOR SERGE Conquered City*
VICTOR SERGE Memoirs of a Revolutionary
VICTOR SERGE Unforgiving Years
SHCHEDRIN The Golovlyov Family
ROBERT SHECKLEY The Store of the Worlds: The Stories of Robert Sheckley*
GEORGES SIMENON Act of Passion*
GEORGES SIMENON Dirty Snow*
GEORGES SIMENON The Engagement
GEORGES SIMENON The Man Who Watched Trains Go By
GEORGES SIMENON Monsieur Monde Vanishes*
GEORGES SIMENON Pedigree*
GEORGES SIMENON Red Lights
GEORGES SIMENON The Strangers in the House
GEORGES SIMENON Three Bedrooms in Manhattan*
GEORGES SIMENON Tropic Moon*
GEORGES SIMENON The Widow*
CHARLES SIMIC Dime-Store Alchemy: The Art of Joseph Cornell
MAY SINCLAIR Mary Olivier: A Life*
TESS SLESINGER The Unpossessed: A Novel of the Thirties
VLADIMIR SOROKIN Ice Trilogy*
VLADIMIR SOROKIN The Queue
DAVID STACTON The Judges of the Secret Court*
JEAN STAFFORD The Mountain Lion
CHRISTINA STEAD Letty Fox: Her Luck

GEORGE R. STEWART Names on the Land
STENDHAL The Life of Henry Brulard
ADALBERT STIFTER Rock Crystal
THEODOR STORM The Rider on the White Horse
JEAN STROUSE Alice James: A Biography*
HOWARD STURGIS Belchamber
ITALO SVEVO As a Man Grows Older
HARVEY SWADOS Nights in the Gardens of Brooklyn
A.J.A. SYMONS The Quest for Corvo
ELIZABETH TAYLOR Angel*
ELIZABETH TAYLOR A Game of Hide and Seek*
HENRY DAVID THOREAU The Journal: 1837–1861*
TATYANA TOLSTAYA The Slynx
TATYANA TOLSTAYA White Walls: Collected Stories
EDWARD JOHN TRELAWNY Records of Shelley, Byron, and the Author
LIONEL TRILLING The Liberal Imagination
LIONEL TRILLING The Middle of the Journey
IVAN TURGENEV Virgin Soil
JULES VALLÈS The Child
MARK VAN DOREN Shakespeare
CARL VAN VECHTEN The Tiger in the House
ELIZABETH VON ARNIM The Enchanted April*
EDWARD LEWIS WALLANT The Tenants of Moonbloom
ROBERT WALSER Berlin Stories*
ROBERT WALSER Jakob von Gunten
REX WARNER Men and Gods
SYLVIA TOWNSEND WARNER Lolly Willowes*
SYLVIA TOWNSEND WARNER Mr. Fortune*
SYLVIA TOWNSEND WARNER Summer Will Show*
ALEKSANDER WAT My Century
C.V. WEDGWOOD The Thirty Years War
SIMONE WEIL AND RACHEL BESPALOFF War and the Iliad
GLENWAY WESCOTT Apartment in Athens*
GLENWAY WESCOTT The Pilgrim Hawk*
REBECCA WEST The Fountain Overflows
EDITH WHARTON The New York Stories of Edith Wharton*
PATRICK WHITE Riders in the Chariot
T.H. WHITE The Goshawk
JOHN WILLIAMS Butcher's Crossing*
JOHN WILLIAMS Stoner*
ANGUS WILSON Anglo-Saxon Attitudes
EDMUND WILSON Memoirs of Hecate County
RUDOLF AND MARGARET WITTKOWER Born Under Saturn
GEOFFREY WOLFF Black Sun*
FRANCIS WYNDHAM The Complete Fiction
JOHN WYNDHAM The Chrysalids
STEFAN ZWEIG Beware of Pity*
STEFAN ZWEIG Chess Story*
STEFAN ZWEIG Confusion
STEFAN ZWEIG Journey Into the Past
STEFAN ZWEIG The Post-Office Girl